The Author

Roger Porkess is Project Leader for Mathematics in Education and Industry (MEI), an independent curriculum development body. In this role he has initiated many recent innovations in school mathematics and statistics and taken a leading role in their subsequent development. He is often invited to contribute to national discussions and to speak at conferences.

Roger read mathematics at Cambridge and later took higher degrees at Reading and Keele. In the early part of his career he taught in secondary schools, both in the U.K. and overseas (Zambia, Ghana and Malaysia). During the early 1980s he was Head of Mathematics at Denstone College in Staffordshire and it was at that time that he became involved in curriculum development work and also became a Collins author.

Since then he has been involved in some eighty books, as author, co-author or series editor, covering the full range of secondary school mathematics and statistics, and also children's plays. He is also heavily involved in school examinations, particularly at A Level.

Roger now lives in Totnes, Devon. He is married with three grown-up children.

Collins

Web-linked Dictionary *of*

Statistics

William Collins' dream of knowledge for all began with the publication of his first book in 1819. A self-educated mill worker, he not only enriched millions of lives, but also founded a flourishing publishing house. Today, staying true to this spirit, Collins books are packed with inspiration, innovation, and practical expertise. They place you at the centre of a world of possibility and give you exactly what you need to explore it.

Collins. Do more.

Collins

Web-linked Dictionary *of* Statistics

Roger Porkess

Collins

An imprint of HarperCollins*Publishers*

HarperCollins*Publishers*
Westerhill Road, Bishopbriggs,
Glasgow G64 2QT

www.collins.co.uk

First published by HarperCollins in 1988
Second edition 2004
Updated 2005

ISBN-10: 0-06-085181-3 (in the United States)
ISBN-13: 978-0-06-085181-1

Originally published in the U.S. copyright © 1991 as
HARPERCOLLINS DICTIONARY OF STATISTICS.

FIRST COLLINS U.S. EDITION 2006
HarperCollins books may be purchased for educational, business, or sales promotional use. For information in the United States, please write to: Special Markets Department, HarperCollins*Publishers*, 10 East 53rd Street, New York, NY 10022.

Typeset by Davidson Pre-Press Graphics Ltd, Glasgow
Printed and bound in Great Britain by Clays Ltd, St Ives plc

05 06 07 08 09 10 9 8 7 6 5 4 3 2 1

To Ketaki

Contents

Preface to the First Edition

In recent years there has been a marked increase in the teaching and application of statistics. A generation ago the subject was rarely taught in schools and very little in universities. Nowadays every school child learns at least some descriptive statistics within the mathematics syllabus; nearly all A-level mathematics courses now contain some probability and statistics topics which for many students, comprise the applied part of their syllabus. In addition, many other subjects, such as geography and biology require some knowledge of statistical techniques from their A-level students. At the tertiary level, the importance of statistical evidence is now accepted by virtually all academic disciplines and there are relatively few science or social science subjects which do not require knowledge of the subject. The increasing use of data bases allows easy access to data and the use of computers takes much of the tedium out of complex calculations.

At the same time public awareness of the subject has increased. For example, it is now expected that when the result of an opinion poll is announced the sample size should be given. Statistical terms, such as parameter and correlation have found their way into everyday English, often used incorrectly.

This rapid growth of the subject means that many people are now using it with a somewhat limited theoretical or mathematical background and it is hoped that such people, be they at work, school or university will find this book particularly helpful.

The context of the Dictionary has been selected to cover the needs not only of those taking statistics as part of A-level, or similar, mathematics courses but also of those learning it as a service subject at the tertiary level. While its format is superficially like that of a traditional dictionary, it differs in a number of respects. Many of the entries are encyclopedic in treatment and include many worked examples. These illustrate how the various statistical measures are calculated and the tests applied. Most of these examples use artificial data; real data is usually more difficult to work with and can easily obscure the point being made. The text is illustrated throughout with diagrams and graphs to aid the reader. Considerable effort has been made to ensure that the information in this book is accessible to its readers. Care has been taken to avoid using terminology that is likely to cause difficulty to the reader of any particular entry. The simpler the subject of an entry, the easier the language and notation used within it. On the other hand, more advanced topics can often only be understood in terms of easier ones and so there is something of a hierarchy among the entries. There is cross-referencing throughout the book, indicated by the use of SMALL CAPITALS.

The conventions used are those in common usage at the time of writing. Greek letters refer to parent population parameters, Roman letters to sample statistics. Capital letters are used for variable names, lower case for values of those variables. A full list of symbols is included in Appendix A while Appendix B is a list of formulae. The tables in Appendix C are those needed for the tests and techniques described within the book.

Acknowledgements

I would like to thank the many people who have helped this book along its way – colleagues, editors, advisors, etc.; in particular Dr Kevin McConway of the Open University and Mr Alan Downie of Imperial College who advised on the text and checked it for accuracy; Janet France for her editorial work on it, including setting up all the cross-referencing; and to Ian Crofton, Edwin Moore and James Carney from Collins Reference Division for their work in turning the typescript into this book.

Roger Porkess

Preface to the Second Edition

The second edition of this book has been the occasion for a major revision of the material. Many of the original entries have been rewritten and extended, and a substantial number of new entries have been included. However, the basic intention of the book remains the same, that it should provide useful information for the large, and increasing, body of users of statistics. That brings with it a word of caution: where possible the entries are accompanied by worked examples and this makes some of them quite long, but readers should not judge the importance of an entry by its length.

I would like to thank Bill Gibson for his help with many of the new entries relating to quality control, and also the members of my family, Ketaki, Sheuli, Veronica and Halley, who have all, in various ways, contributed to this book.

Roger Porkess
May 2004

a

abscissa the horizontal or x-coordinate in a two-dimensional system of cartesian coordinates. The vertical or y-coordinate is called the *ordinate*. See Fig. 1.

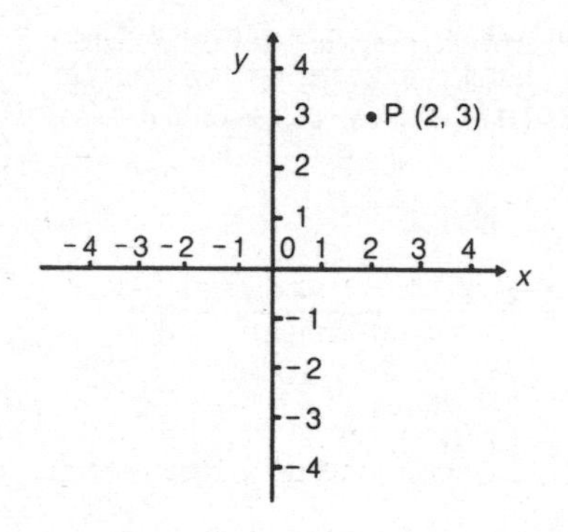

Fig. 1. **Abscissa.** In this example, the abscissa of P is 2, and the ordinate of P is 3.

absolute frequency see FREQUENCY.

absolute measure of dispersion a measure of dispersion (or spread) relative to the size and distribution of the figures involved.

Example: The sets of numbers P and Q

P	2	4	6	8	10	12	14	16
Q	1002	1004	1006	1008	1010	1012	1014	1016

both have the same range (14), interquartile range (8) and standard deviation (4.9). Relative to the sizes of the numbers, however, P is clearly much more spread than Q. If, for example, the figures represented typical incomes of people in two countries (in suitable units of currency), they would be remarkably uniform in Q but not at all so in P.

There are several absolute measures.

(a) *Coefficient of variation*. This is given by:

$$\frac{\text{standard deviation}}{\text{mean}}$$

In this example, the values are P: $\frac{4.9}{9} = 0.54$; Q: $\frac{4.9}{1009} = 0.0049$

Coefficient of variation is sometimes written as a percentage:

$$\frac{\text{standard deviation}}{\text{mean}} \times 100\%$$

(b) *Quartile coefficient of dispersion*. This is given by

$$\frac{\text{upper quartile} - \text{lower quartile}}{\text{upper quartile} + \text{lower quartile}}$$

In this example, the values are P: $\frac{8}{18} = 0.44$; Q: $\frac{8}{2018} = 0.0040$

(c) *Coefficient of mean deviation*. This is given by:

$$\frac{\text{mean absolute deviation}}{\text{mean}}$$

In this example, the values are P: $\frac{4}{9} = 0.44$; Q: $\frac{4}{1009} = 0.0040$

A similar coefficient can be found using the median or mode instead of the mean.

absolute value or **modulus** the magnitude of a real number or quantity, regardless of its sign. Symbol: straight brackets $| \, |$.

$$|-6| = 6 \qquad |17.3| = 17.3$$

absorbing barrier a state for a random walk such that, if the particle making the walk ever reaches it, the random walk ceases.

Example: two gamblers, A and B, take turns in a game. At each turn, one gambler wins some money from the other, with some chance mechanism deciding who wins. Suppose that A starts with a units of capital, and B with b. Let the random walk X_n represent the cumulative gain of (say) B after n turns. If at any stage $X_n = a$, then B has gained all A's capital, A is ruined and the game must stop. Similarly, if at any stage $X_n = -b$, then B has no money left and the game stops. The random walk X_n has absorbing barriers at a and $-b$. (This is the classical "gambler's ruin" problem; it has been generalised and extended in many ways, and applied to many problems in business, finance and commerce.)

acceptance number or **allowable defects** the greatest number of defectives in a sample that will allow the batch (from which the sample is drawn) to be accepted without further quality-control testing. In a single sampling scheme for quality control, a sample is taken from each batch of the product. If the number of defectives is no larger than the acceptance number, the whole batch is accepted.

acceptance region the set of values of a test statistic for which the null hypothesis is accepted. See NULL HYPOTHESIS, CRITICAL REGION.

acceptance sampling a term used in quality control to describe the sampling process used before accepting a batch of goods from a supplier.

A manufacturer usually uses components or other materials from someone else as the input to the process. If these incoming goods are not of a satisfactory standard, the final product is likely to be substandard and unsatisfactory. The supplier of these incoming goods could be a separate firm or a different section of the same firm. To try to guard against problems from poor incoming goods, a manufacturer inspects them before accepting them. This inspection can range from a cursory glance to checking every item. A compromise between these two extremes is to take a random sample from the batch of incoming goods and check these items. If these are up to standard, then the whole batch is accepted, if they are not then the whole batch is rejected. The rule that is used to decide whether to accept a batch is called the *decision rule*. See ACCEPTANCE NUMBER.

addition law a law concerning the probabilities of two events occurring. The addition law states that:

$P(A \text{ or } B) = P(A) + P(B) - P(A \text{ and } B)$

It is often written, using set notation, as:

$P(A \cup B) = P(A) + P(B) - P(A \cap B)$

Example: A card is drawn at random from a pack of 52 playing cards. What is the probability that it is a club or a queen? (See Fig. 2).

Event A: The card is a club. Probability = 1/4

Event B: The card is a queen. Probability = 1/13

Event $A \cap B$: The card is the queen of clubs. Probability = 1/52

Event $A \cup B$: The card is a club or a queen.

$$P(A \cup B) = P(A) + P(B) - P(A \cap B)$$
$$= 1/4 + 1/13 - 1/52 = 16/52 = 4/13$$

The probability that the card is a club or a queen is 4/13.

Event B

	2	3	4	5	6	7	8	9	10	J	Q	K	A
♠	×	×	×	×	×	×	×	×	×	×	×	×	×
♥	×	×	×	×	×	×	×	×	×	×	×	×	×
♦	×	×	×	×	×	×	×	×	×	×	×	×	×
Event A → ♣	×	×	×	×	×	×	×	×	×	×	×	×	×

Event $A \cap B$

Fig. 2. **Addition law.** Picking a card at random from the pack. Event $A \cup B$ occurs when either a club or queen is drawn. The probability of Event $A \cup B$ is 4/13.

additive model a model in which individual terms are added together.

age **1.** the period of time that a person, animal or plant has lived, or that an object has existed, i.e. the *chronological age*. In conventional use, a person's age is often given as the number of completed years; in that case it is a discrete variable taking integer values.

2. the level in years that a person has reached in any area of development on a scale determined by the normal level of achievement for that chronological age, for example, *mental age*, *reading age*.

aggregrate the value of a single variable resulting from the combination of data for a number of variables. For example, the cost-of-living index is an aggregate of the various components that are used to form it. An examination mark is often an aggregate of the marks of several different papers, perhaps also including a project. The process of forming an aggregate is called *aggregation*.

alienation see COEFFICIENT OF ALIENATION.

allowable defects see ACCEPTANCE NUMBER.

alternative hypothesis see NULL HYPOTHESIS.

analysis of variance, ANOVA a widely used technique for comparing the means of several populations, given samples of observations from those populations. It is based on an analysis of the total variation displayed by the data, splitting this into variation between the samples and variation within the samples, and then comparing these components. For the technique to be valid, the populations need to be Normal with equal variances. The technique can still be a good approximation if these assumptions are not correct.

The idea underlying one way analysis of variance is illustrated in Fig. 3. Each of the two diagrams illustrates three samples taken to investigate whether the means of the three populations from which they were taken may be assumed to be equal. The vertical lines on the horizontal axis represent the means of each of the three samples. It can be seen that the means of the two *x*-samples are the same; similarly for the two *y*-samples and the two *z*-samples.

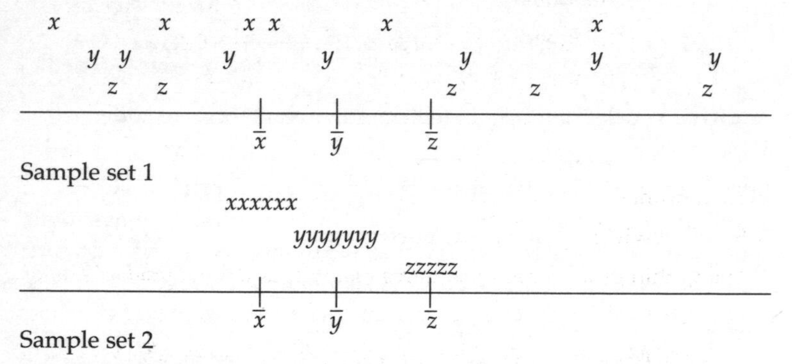

Fig. 3. **Analysis of variance.** Illustration of variation between samples and within samples.

Although the means of the *x*-, *y*- and *z*- samples are the same in each case, an observer of the second set would be much more confident that the three population means are different than an observer of the first set. In the second set of samples, the variation within each sample is small compared to that within the whole group, whereas for the first set the variation within the samples is comparable to that within the whole group.

In analysis of variance, the variance is estimated in two different ways. The first way does not assume that the means of the populations are equal, but just looks at the individual samples and then pools the estimates obtained; this is called the "within groups" estimate. The second way does assume that the means of the populations are equal and therefore treats the observations as though they all came from one sample; this is the "between groups" estimate.

In the first set of samples in Fig. 3, the two methods would give similar estimates of the variance. In the second set, the within groups estimate would be much smaller than the between groups estimate.

Analysis of variance provides a 1-tail test of the null hypothesis that the means of the treatments are equal against the alternative hypothesis that the means of the treatments are not all equal. The ratio of these two variance estimates has an *F* distribution if this null hypothesis is true. The test statistic, *F*, is calculated as follows.

Suppose that there are *k* treatments (i=1, ...,k) and that treatment i has n_i observations. x_{ij} is the jth observation on the ith treatment.

Treatment 1	$x_{1\,1}, x_{1\,2}, \dots x_{1\,n_1}$	Mean $\bar{x}_1$
Treatment 2	$x_{2\,1}, x_{2\,2}, \dots x_{2\,n_2}$	Mean $\bar{x}_2$
Treatment 3	$x_{3\,1}, x_{3\,2}, \dots x_{3\,n_3}$	Mean $\bar{x}_3$
....	....	....
Treatment *k*	$x_{k\,1}, x_{k\,2}, \dots x_{k\,n_k}$	Mean $\bar{x}_k$

Total number of observations, $n = n_1 + n_2 + \dots + n_k$

For the "within groups" variance

The within groups sum of squares, usually called the *residual sum of squares*

$$SS_R = \sum_{j=1}^{n_1}\left(x_{1j} - \bar{x}_1\right)^2 + \sum_{j=1}^{n_2}\left(x_{2j} - \bar{x}_2\right)^2 + \dots + \sum_{j=1}^{n_k}\left(x_{kj} - \bar{x}_k\right)^2 = \sum_{i=1}^{k}\sum_{j=1}^{n_i}\left(x_{ij} - \bar{x}_i\right)^2$$

In practice, of course, computers are usually used to do these calculations.

There are $n - k$ degrees of freedom for the within group sum of squares.

The within groups mean square, usually called the residual mean square, is now calculated as

$$MS_R = \frac{SS_R}{n-k}\,.$$

This is the within groups estimate of variance.

For the "between groups" variance

$$\text{Overall mean}, \ \bar{x} = \frac{n_1\bar{x}_1 + n_2\bar{x}_2 + \dots + n_k\bar{x}_k}{n} = \frac{\sum_{i=1}^{k}\sum_{j=1}^{n_i} x_{ij}}{n}$$

The between groups sum of squares,

$$SS_B = n_1(\bar{x}_1 - \bar{x})^2 + n_2(\bar{x}_2 - \bar{x})^2 + \ldots + n_k(\bar{x}_k - \bar{x})^2.$$

Again the actual calculation is usually done by a computer package.

There are $k - 1$ degrees of freedom for the between groups sum of squares.

The between groups mean square is now calculated as

$$MS_B = \frac{SS_B}{k-1}$$

This is the between groups estimate of variance.

The test statistic, F, is calculated as $F = \frac{MS_B}{MS_R}$ and this is tested by reference to the F distribution with $k - 1$, $n - k$ degrees of freedom, i.e. $F_{k-1,n-k}$.

Example: There are five treatments for lowering blood pressure. An initial test is needed of whether there is any real difference between them. Each treatment is given to a different randomly chosen sample of people with high blood pressure. The results, using suitable units, are as follows.

H_0 the means of the treatments are equal.

H_1 the means of the treatments are not all equal.

1-tail test

5% significance level

Treatment	Results						
1	12	6	5	7	10		
2	10	15	14	13	12	12	15
3	3	2	7	8	3	1	
4	7	8	7	10			
5	16	18	21	19	21		

For these data, $k = 5$

$n_1 = 5$, $n_2 = 7$, $n_3 = 6$, $n_4 = 4$, $n_5 = 5$ and $n = 27$

The treatment means are

$\bar{x}_1 = 8$, $\bar{x}_2 = 13$, $\bar{x}_3 = 4$, $\bar{x}_4 = 8$, $\bar{x}_5 = 19$

and the overall mean is $\bar{x} = 10.44$

Within groups

$$\text{Sum of squares, } SS_R = 370$$
$$\text{Degrees of freedom } n - k = 22$$
$$\text{Mean square } MS_R = \frac{370}{27-5} = 16.82$$

Between groups

$$\text{Sum of squares, } SS_B = 738.6$$
$$\text{Degrees of freedom } k - 1 = 4$$
$$\text{Mean square } MS_B = \frac{736.6}{5-1} = 184.6$$

The test statistic $F = \dfrac{184.6}{16.82} = 11.0$.

The critical value for $F_{4,22}$ at the 5% significance level is 2.82 (see Table 5).

Since 11.0>2.82, the null hypothesis is rejected. The different treatments do not have the same mean effect.

The technique of analysis of variance can readily be extended to situations where more than one factor might influence the outcome of an experiment (for example in factorial experiments). In all cases, appropriate mean squares are calculated, by dividing sums of squares by degrees of freedom. The mean squares are then compared with each other using *F* tests.

Other situations, such as regression, are often analysed using an analysis of variance format.

anova *abbrev. for* ANALYSIS OF VARIANCE.

antilogarithm see LOGARITHM.

AOQ see AVERAGE OUTGOING QUALITY.

AOQL see AVERAGE OUTGOING QUALITY LIMIT.

approximation the process or result of making a rough calculation, estimate or guess. Approximation is used extensively when deciding whether an answer is reasonable. The statement 'It takes five hours to fly from Britain to Australia' is clearly false. The distance is about 20,000 km, the speed about 1,000 km per hour, and so the time taken is about 20 hours. To decide that the statement was false, it was not necessary to know exact values of the speed of the aeroplane or the distance involved.

When giving numerical information, the number of significant figures should be consistent with the accuracy of the information involved. Thus the statement 'The cost of running my car is £1,783.47 per year' is misleading because it is not possible to estimate it so accurately, especially when depreciation is taken into account. A better figure would have been 'approximately £2,000', which is obviously a round number. See DECIMAL PLACES, ROUNDING, SIGNIFICANT FIGURES.

arithmetic mean the result obtained by adding the numbers or quantities in a set and dividing the total by the number of members in the set. For example, the arithmetic mean of 43, 49, 63, 51 and 28 is:

$$\frac{43 + 49 + 63 + 51 + 28}{5} = 46.8$$

The arithmetic mean of the numbers $x_1, x_2, \ldots x_n$ is denoted by $\bar{x}$ and given by:

$$\bar{x} = \frac{x_1 + x_2 + \ldots + x_n}{n}$$

This may also be written as:

$$\bar{x} = \frac{1}{n}\sum_{i=1}^{n} x_i \quad \text{or as} \quad \bar{x} = \frac{\sum_{i=1}^{k} x_i f_i}{\sum_{i=1}^{k} f_i}$$

where f_i is the frequency of the value x_i, and there are k distinct values of x.

Arithmetic mean is often referred to as MEAN, and also as AVERAGE.

assignable cause or special cause a term used in quality control for the cause of a machine or process going wrong.

association the tendency of two events to occur together. Suppose the probabilities (or frequencies) of the different combinations of events *A* and not *A*, and *B* and not *B*, are

	A	not *A*
B	*a*	*b*
not *B*	*c*	*d*

The association is said to be positive if $ad > bc$, and negative if $ad < bc$. If $ad = bc$, the events *A* and *B* are independent. *Yule's coefficient of association* is given by:

$$\frac{ad - bc}{ad + bc}$$

The term association is also applied to variables, X and Y (say). Some pairs of variables tend to change their values together.

In some cases large values of X tend to occur with large values of Y, and small values of X with small values of Y. An example of this is when X is the heights of people and Y their weights. Such variables are said to be *positively associated*, but in this example the association is of course far from perfect. Variables are *negatively associated* when large values of X tend to occur with small values of Y, and vice versa. If there is no such tendency at all, the variables are independent.

Association can be measured by covariance or correlation when the relationship between the variables is linear. The term correlation is often used somewhat loosely to be synonymous with association but this is incorrect as it does not cover non-linear association.

assumed mean an estimated or approximate value for the ARITHMETIC MEAN, or average, used to simplify its calculation. When working out the arithmetic mean of a set of numbers of similar size, the calculation can sometimes be simplified by using an assumed mean. This is shown in the following example.

Example: find the mean of the ages of six children whose ages, in years and months, are 13y 11m, 14y 4m, 13y 6m, 14y 6m, 14y 1m and 14y 2m,

(a) Take an assumed mean: 14y 0m;

(b) Calculate the differences from the assumed mean:
-1, +4, -6, +6, +1 and +2 months;

(c) Calculate the mean of the difference:

$$\frac{-1+4-6+6+1+2}{6} = +1 \text{ month}$$

(d) The mean age of the children is given by:
true mean = assumed mean + mean of differences
= 14y 0m + 1m

The mean age of the children is 14y 1m

It does not matter what value is taken for the assumed mean, but the nearer it is to the true mean, the smaller are the numbers involved in the calculation.

attribute testing testing a product in quality control when the result is either 'good' or 'defective'. Attribute testing may be contrasted with variable testing where the result is a quantitative measure, like the length of a nail or the resistance of an electrical component. See also LATTICE DIAGRAM.

autocorrelation a measure associated with time series.

When considering using time series data for forecasting, it is important to know whether the current and earlier values give any information about subsequent values. In autocorrelation, pairs of data values from different times are regarded as points in a bivariate distribution. If this bivariate distribution shows a high level of correlation, then it is reasonable to use existing data for forecasting.

Example 1: the data in the table below are the sales figures for a new product. Is it reasonable to use them for forecasting?

Month	Jan	Feb	Mar	Apr	May	June	July	Aug
Sales × £100	35	36	39	39	41	45	45	42

There are 8 data values here and putting successive values together in pairs gives the following 7 bivariate data points:

(35, 36), (36, 39), (39, 39), (39, 41), (41, 45), (45, 45) and (45, 42).

When these points are plotted on a scatter diagram (Fig. 4), they lie close to a straight line suggesting positive correlation. The correlation coefficient is actually 0.82. With such a high value, it is indeed reasonable to use existing data to forecast future sales. The data are said to be autocorrelated because they are correlated with themselves.

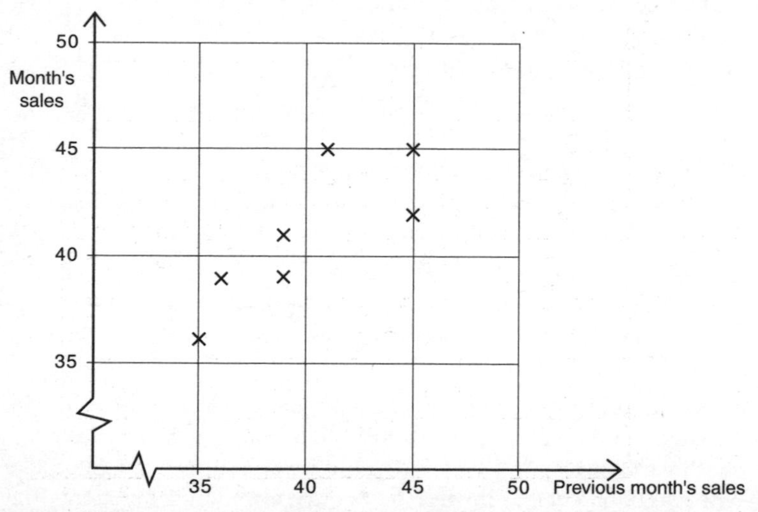

Fig. 4. **Autocorrelation.** A time series plotted as bivariate data on a scatter diagram.

In the example above, the correlation was carried out with a lag of 1 time unit; successive values were used to form the bivariate points. It is, however, essential to consider other time lags if a proper understanding is to be built up. This is illustrated in the next example where the data are quarterly and show a marked seasonal component.

Example 2: the quarterly production figures for a particular product over a 5 year period are given below.

	1st quarter	2nd quarter	3rd quarter	4th quarter
Year 1	252	279	313	249
Year 2	266	290	290	271
Year 3	249	249	299	283
Year 4	233	272	285	275
Year 5	271	280	311	268

These figures are illustrated in Fig. 5.

If autocorrelation is carried out on these figures with a lag of 1 time unit, the correlation coefficient is $r = -0.077$. There is no correlation with lag 1.

However, the autocorrelation can be carried out with a lag of 2 time units, so that the bivariate points are:

(252, 313), (279, 249) … . In this case there is strong negative correlation with $r = -0.622$.

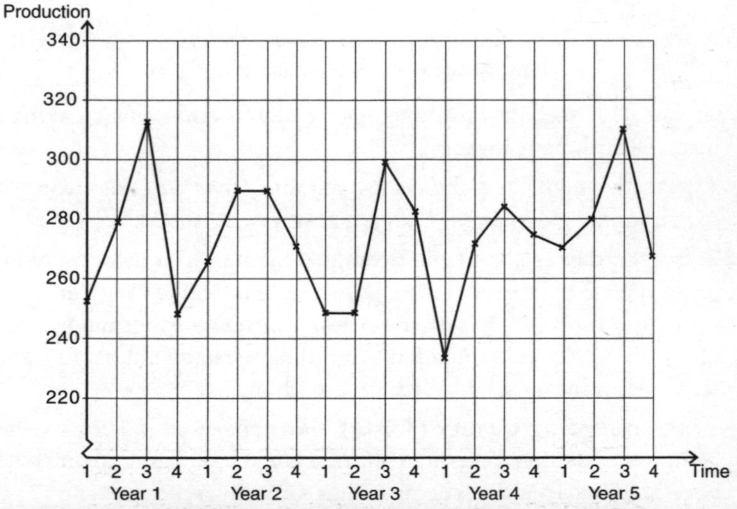

Fig. 5. **Autocorrelation.** Quarterly data for 5 years' production.

With a lag of 4 time units, the first two points are (252, 266) and (279, 290) and in this case there is positive correlation with $r = 0.471$.

The reason that this occurs can be seen from the data or from the plot. The high values usually occur in the 3rd quarter, the low values in the 1st and 4th quarters. A lag of 4 time units associates the high values of one year with those from the previous year, and similarly for the low values. Consequently they are correlated. By contrast a lag of 2 time units associates high values with low values and so there is negative correlation.

In Fig. 2(d) the correlation coefficients are plotted against the lag. This type of diagram is called an *autocorrelogram* or a *correlogram*. It highlights the cyclical nature of the data, and the period of the cycle.

autocorrelogram see AUTOCORRELATION.

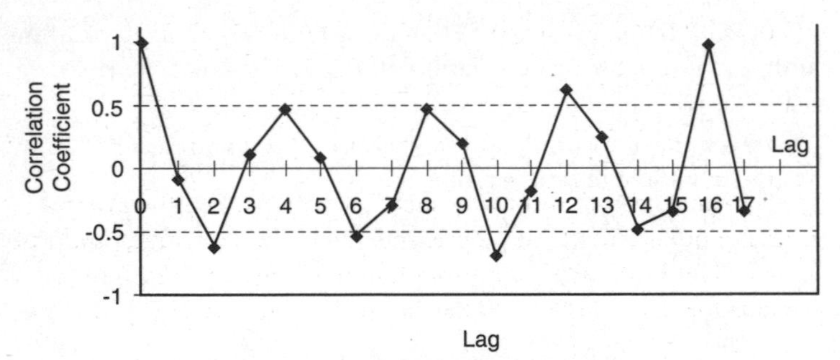

Fig. 6. **Autocorrelation.** An autocorrelogram.

average **1.** in technical use, average usually has the same meaning as mean or arithmetic mean.

2. In certain contexts technical use requires other types of mean. Examples are given under GEOMETRIC MEAN and HARMONIC MEAN.

3. In everyday use, the word average is often used more loosely to mean typical or representative, as in a statement like "William is average at football". It is often unclear which, if any, formal measure of central tendency is intended. According to context, it may be any (or none) of mean, mode, median or midrange.

average outgoing quality (AOQ) the proportion of defective items being sent out after a quality control scheme has been implemented. For a single inspection scheme (single sampling scheme) with acceptance number n from samples of size N, with 100% inspection of failed batches, AOQ is given by

$$\mathrm{AOQ} = \mathrm{P}(p)\,(1- \frac{n}{N})p$$

where p is the proportion of defective items manufactured, and $\mathrm{P}(p)$ is the probability of a batch with proportion p defective being accepted.

average outgoing quality limit (AOQL) the highest average proportion of defective items that will be let through in the long run, following a quality-control sampling scheme.

When average outgoing quality (AOQ) is plotted against the proportion of defectives manufactured, p, the graph rises to a maximum and then falls away. A typical graph for a single sampling scheme is shown in Fig. 7. If p is small, the AOQ will be low because there are few defectives to get through. If p is large, the sampling scheme is likely to result in most batches being stopped and undergoing 100% inspection; any defectives are then rejected. Between these two cases are values of p which are neither very small, nor large enough to involve most batches being stopped.

The maximum value shown on such a graph is called the average outgoing quality limit, and represents the highest average proportion of defectives that will be let through in the long run. It is, however possible for worse individual batches to get through.

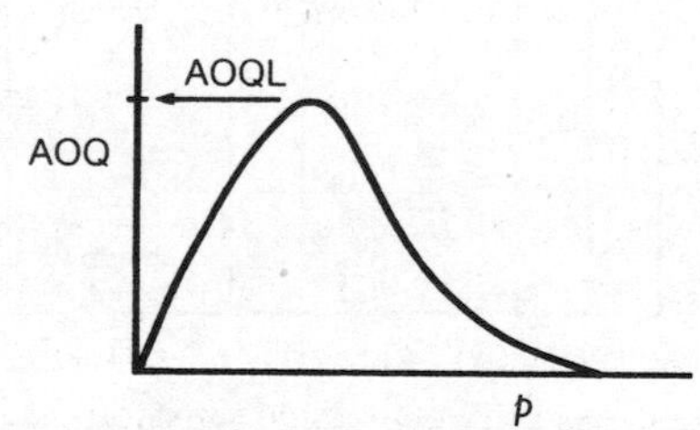

Fig. 7. **Average outgoing quality limit (AOQL).** The average outgoing quality (AOQ) of goods is plotted against the proportion of defective goods (p) to find the maximum value (AOQL).

b

band chart a percentage bar chart.

bar chart or **bar diagram** a data display using a number of rectangles (bars), of the same width, with lengths proportional to the frequencies they represent (Fig. 8). The bars may be drawn vertically or horizontally. It is customary to leave a fixed gap between them. A *component* or *compound bar chart* has two or more parts to each bar (Fig. 9). Bar charts are most commonly used to represent *categorical data*. There are many ways in which such diagrams can be drawn, some of them very artistic (Fig. 10). See also FREQUENCY CHART, HISTOGRAM, POPULATION PROFILE, THREE-QUARTERS HIGH RULE. Compare VERTICAL LINE CHART.

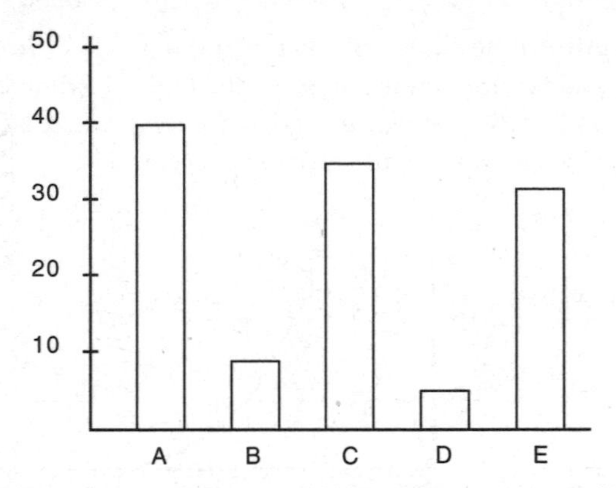

Fig. 8. **Bar chart.** Results of a survey of 120 people into the best of 5 possible new flavours of yoghurt, A, B, C, D and E.

bar diagram see BAR CHART.

batch a collection of items, usually with some common feature. For example, if the quality of components being produced by a factory is being monitored, a batch might be the output from one particular shift.

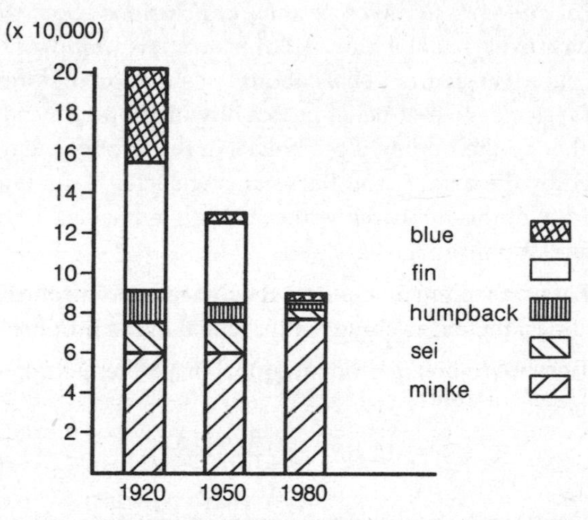

Fig. 9. **Bar chart.** Compound bar chart of estimated population of five types of whale in an area of the Southern Ocean.

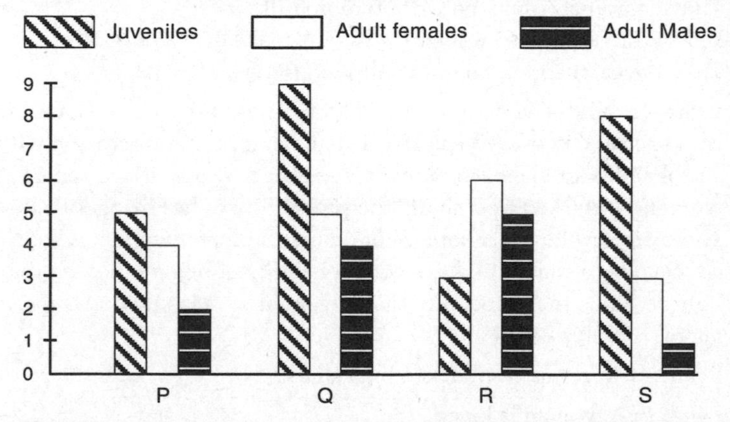

Fig. 10. **Bar chart.** The number of juveniles, adult females and adult males in 4 colonies, P, Q, R and S of a small rodent.

Bayesian controversy a controversy between two schools of statistical thought. In Bayesian statistics, probability is thought of as a numerical degree of belief. In some situations, like throwing a pair of dice, the belief can be supported objectively. In others, such as the predicted price of gold next week, any probability used is essentially subjective. In non-Bayesian statistics, probabilities rely on a frequency-based interpretation.

Statistical inference involves drawing conclusions about the situation underlying experimental data. A Bayesian statistician will often do this by first expressing prior beliefs about the situation (before collecting any data) as a degree-of-belief probability distribution, and then using Bayes' theorem to update those beliefs in the light of the information provided by the data. A non-Bayesian statistician would not accept the validity of this approach without a frequency-based interpretation of the probabilities.

[Note: Bayes' theorem is merely a result concerning conditional probabilities; there is nothing controversial about it in itself.]

Bayes' theorem (on conditional probability), the theorem which states that

$$P(A|B) = \frac{P(B|A)P(A)}{P(B)}$$

where $P(B|A)$ is the probability of event B occurring, given that event A has occurred.

The denominator may be written more fully using $P(B)=P(B|A)P(A)+P(B|A')P(A')$, where A' is the event not–A. Thus Bayes' theorem relates $P(A|B)$ to $P(B|A)$ and $P(B|A')$.

Example: all the women from a Polynesian island go on a shark hunt on a special day every year. It is a somewhat perilous occasion with a probability of $\frac{1}{10}$ that a woman is attacked by a shark; when a woman is attacked by a shark, the probability of her being killed is $\frac{1}{3}$. There are several other sources of danger, so that a woman who is not attacked by a shark still has a $\frac{1}{60}$ probability of being killed on the hunt. What is the probability that a woman who is killed was the victim of a shark?

Event A: A woman is attacked by a shark.

Event B: A woman is killed.

$$P(A) = \frac{1}{10} \qquad P(B|A) = \frac{1}{3}$$

$$P(A') = \frac{9}{10} \qquad P(B|A') = \frac{1}{60}$$

Then Bayes' theorem gives:

$$P(B|A) = \frac{\frac{1}{3} \times \frac{1}{10}}{\frac{1}{3} \times \frac{1}{10} + \frac{1}{60} \times \frac{9}{10}} = \frac{20}{29}$$

So the probability that a woman who is killed was the victim of a shark is $\frac{20}{29}$.

Bayes' theorem can be written in more general form as

$$P(A_r|B) = \frac{P(B|A_r) \times P(A_r)}{P(B)}$$

and

$$P(B) = \sum_{r=1}^{r=n} P(B|A_r) \times P(A_r)$$

where $A_1, A_2, \ldots A_n$ are mutually exclusive and exhaustive events.

Bernoulli distribution the special case of the binomial distribution when $n = 1$.

Number of successes, r:	0	1
Probability, P(X=*r*):	$1\text{-}p$	p

Thus it is the probability distribution of the number of successes in a single trial.

Bernoulli's theorem the theorem which states that, if the probability of success in a trial is p, the probability of exactly r successes in n independent trials is:

$$\binom{n}{r} p^r(1-p)^{n-r} \text{ where } \binom{n}{r} = \frac{n!}{(n-r)!r!}$$

(See also BINOMIAL COEFFICIENTS.) This is the theorem underlying the binomial distribution.

Bernoulli trial an experiment with fixed probability p of success, $1 - p$ of failure. In a sequence of Bernoulli trials, the probability of success remains constant from one trial to the next, independent of the outcome of previous trials.

The probability distribution for the various possible numbers of successes, 0 to n, in n Bernoulli trials is the binomial distribution. In the case when $n = 1$, this can also be called the Bernoulli distribution.

beta distribution the distribution with probability density function given, for $0 \leqslant x \leqslant 1$, by:

$$f(x) = \frac{x^{a-1}(1-x)^{b-1}}{B(a,b)}.$$

There are two parameters a and b, $a > 0$ and $b > 0$, and $B(a,b)$ is the *beta function* given by:

$$B(a,b) = \int_0^1 x^{a-1}(1-x)^{b-1}dx.$$

The beta distribution has

$$Mean = \frac{a}{(a+b)},\ Variance = \frac{ab}{(a+b+1)(a+b)}$$

$$\text{and } Mode = \frac{a-1}{a+b-2} \text{ (provided } a > 1,\ b > 1).$$

The graph of the beta distribution varies in shape according to the values of a and b, as shown in Fig. 11. This means that the beta distribution provides a very flexible model that is often suitable for variables constrained to lie in the interval (0,1) (or in any other finite interval with suitable scaling). For example a beta distribution with a and b only slightly greater than 0 might provide a suitable degrees-of-belief prior distribution (see Bayesian controversy) for the proportion of defective items in a batch where it is thought that batches are likely to be either of very good quality or (perhaps because of a major fault in the process) of very poor quality.

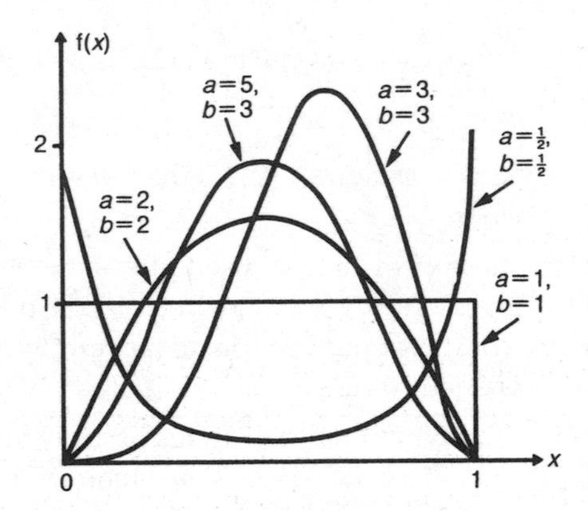

Fig. 11. **Beta distribution.**

beta function see BETA DISTRIBUTION.

bias a systematic error in a statistical result. Errors from chance will cancel each other out in the long run, but those from bias will not.

Example: the police decide to estimate the average speed of drivers using the fast lane of a motorway, and consider how it can be done. One method suggested is to tail cars, using a police patrol car, and record their speeds as being the same as that of the police car. This is likely to produce a biased estimate, as any driver exceeding the speed limit will slow down on seeing a police car coming up behind; so the estimate will be too low. To eliminate this source of bias, the police could use an unmarked car.

biased estimator a formula or procedure which produces a biased estimate of a parameter of the parent population from sample data. In statistics it is common to estimate properties of a parent population, like its mean and variance, by examining a sample. A formula or procedure which predicts a property or parameter of the parent population using the data is called an estimator. If its expected value over all possible samples is equal to the quantity it estimates, an estimator is unbiased. Otherwise it is biased.

A sample of size n is taken from a population. The values of a random variable associated with the items in the sample are $x_1, x_2, \ldots, x_n$.

The sample mean,

$$\bar{x} = \frac{x_1 + x_2 + \ldots + x_n}{n}$$

is an *unbiased estimator* for the population mean. An unbiased estimator for the population variance is

$$\sum_{i=1}^{n} \frac{(x_i - \bar{x})^2}{n-1}$$

By contrast the quantity

$$\sum_{i=1}^{n} \frac{(x_i - \bar{x})^2}{n}$$

is a biased estimator for the population variance.

Individual values of an unbiased estimator may be a long way away from the parameter being estimated but on average it gets it right. For a biased estimator, individual values may be close to that of the parameter, but on average it gets it wrong.

bimodal (of a distribution) a distribution which has two distinct, local modes, is described as bimodal. Strictly the modes should be equal, but in practice the term is often applied to distributions, like that in Fig. 12; the two peaks on the graph are unequal in height but comparable and clearly higher than the rest of the distribution.

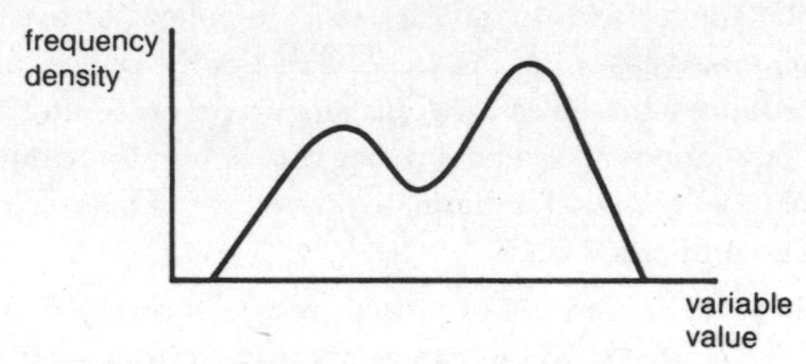

Fig. 12. **Bimodal.** A bimodal distribution.

The term is also applied to frequencies where it is common to be stricter about them being equal, as in the example below.

Example: the goals scored per match by a football team in a season are given in the table below. There are two modes, 0 and 2, so the distribution is bimodal.

Goals/match	*Frequency (matches)*
0	12
1	8
2	12
3	6
4	3
5	1

binomial coefficients the coefficients of the terms in the expansion of $(x + y)^n$ for positive integer values of n. Notation: $\binom{n}{r}$ or ${}^nC_r$.

Example:

$$(x+y)^5 = 1x^5 + 5x^4y + 10x^3y^2 + 10x^2y^3 + 5xy^4 + 1y^5$$

and so the binomial coefficients for $n = 5$ are 1, 5, 10, 10, 5, 1. There are two common notations for writing these numbers:

(a) $\binom{5}{0}, \binom{5}{1}, \binom{5}{2}, \binom{5}{3}, \binom{5}{4}$ and $\binom{5}{5}$,

(b) ${}^5C_0$, ${}^5C_1$, ${}^5C_2$, ${}^5C_3$, and ${}^5C_5$.

There are two commonly used ways of working out binomial coefficients.

(a) Use of the formula $\binom{n}{r}=\frac{n!}{r!(n-r)!}$.

Thus $\binom{5}{2}=\frac{5!}{2!3!}=\frac{5.4.3.2.1}{2.1.3.2.1}=10$.

(b) Pascal's triangle:

$$
\begin{array}{ccccccccccccccc}
 & & & & & & & 1 & & & & & & & \\
 & & & & & & 1 & & 1 & & & & & & \\
 & & & & & 1 & & 2 & & 1 & & & & & \\
 & & & & 1 & & 3 & & 3 & & 1 & & & & \\
 & & & 1 & & 4 & & 6 & & 4 & & 1 & & & \\
 & & 1 & & 5 & & 10 & & 10 & & 5 & & 1 & & \\
 & 1 & & 6 & & 15 & & 20 & & 15 & & 6 & & 1 & \\
\bullet & & \bullet & & \bullet & & \bullet & & \bullet & & \bullet & & \bullet & & \bullet
\end{array}
$$

Apart from the outside numbers, all of them 1, each coefficient in a new row is found by adding the two numbers above it together.

Relationships involving binomial coefficients include:

Symmetry: $\binom{n}{r}=\binom{n}{n-r}$

Going from one row to the next: $\binom{n}{r}+\binom{n}{r+1}=\binom{n+1}{r+1}$

Total in any row: $\binom{n}{0}+\binom{n}{1}+\binom{n}{2}+\ldots+\binom{n}{n}=2^n$

The number of ways of selecting r objects from n, without regard to their order is given by:

$$\binom{n}{r}$$

binomial distribution the probability distribution of the various possible numbers of successes (0 to n) when n independent Bernoulli trials are carried out. Notation: $B(n,p)$. See also BERNOULLI DISTRIBUTION. Compare PASCAL'S DISTRIBUTION.

If the probability of success is denoted by p, and that of failure by $q = 1 - p$, then the probability of r successes and $n - r$ failures is given by:

$\binom{n}{r} p^r q^{n-r} = \binom{n}{r} p^r (1-p)^{n-r}$, where the binominal coefficient $\binom{n}{r} = \frac{n!}{r!(n-r)!}$.

The mean, or expectation, of the binomial distribution is given by np, the standard deviation by $\sqrt{npq}$.

Example: in a certain country, 60% of people support the Democratic Party, the rest the Parliamentary Party.

(a) Five people are selected at random. What are the probabilities that 0, 1, 2, 3, 4, and 5 of them support the Parliamentary Party?

(b) Opinion pollsters interview people in sets of five. What are the mean and standard deviation of the numbers of Parliamentary Party supporters in these sets?

(a)

Number supporting the Parliamentary Party		*Probability*
0	$\binom{5}{0}(0.6)^5$	0.0778
1	$\binom{5}{1}(0.6)^4(0.4)$	0.2592
2	$\binom{5}{2}(0.6)^3(0.4)^2$	0.3456
3	$\binom{5}{3}(0.6)^2(0.4)^3$	0.2304
4	$\binom{5}{4}(0.6)(0.4)^4$	0.0768
5	$\binom{5}{5}(0.4)^5$	0.0102

(b) *Mean* $= np = 5 \times 0.4 = 2$

Standard deviation $= \sqrt{npq} = \sqrt{5 \times 0.4 \times 0.6} = 1.095$

The binomial distribution can be approximated by:

- The Normal distribution, mean np and standard deviation $\sqrt{npq}$, when n is large and p is neither small nor near 1.

- The Poisson distribution, mean and variance both np, when n is large and p small.

The distribution of the proportion of successes in n trials is binomial with

$$\textit{Mean} = p$$
$$\textit{Standard deviation} = \sqrt{\frac{pq}{n}}$$

The probability generating function (see PROBABILITY GENERATING FUNCTION) for the binomial distribution is given by $(q + pt)^n$ since, when expanded this gives:

$$\binom{n}{0}q^n + \binom{n}{1}q^{n-1}pt + \ldots + \binom{n}{r}q^{n-r}p^r t^r + \ldots + \binom{n}{n}p^n t^n.$$

biometry the study of biological data by means of statistical analysis.

birth and death process a stochastic process representing the evolution of a population where births and deaths occur. It is sometimes generalised to allow immigration and/or emigration. The idea is applied to many phenomena, not just human or animal populations.

birth rate the ratio of live births per unit time in a specified area, group, etc., to population, usually expressed per thousand population per year. In the 1870s, the birth rate in the UK was about 35 per 1000, compared to 11.0 in 2003, the lowest recorded figure.

biserial (of the relationship between two variables) having one variable that only takes two values.

bivariate distribution a distribution of two random variables. Compare MULTIVARIATE DISTRIBUTION.

Example: a second-hand car, of a particular make and model, has two easily measurable variables of interest to a prospective buyer: its age and the mileage it has covered. (There are also a number of other factors, like its general condition and the driving habits of the previous owner.) A dealer, buying a batch of 30 cars of a particular model from a company which runs a fleet of them, tabulates the information as a bivariate frequency or contingency table, as in Fig. 13.

In this example the two variables, age and mileage, are not independent; the older cars have usually covered a greater distance. It would,

		MILEAGE (x 1000 miles)							
		0–	10–	20–	30–	40–	50–	60–	70–
AGE (years)	0–	1				1	1		
	1–					2			
	2–		1	1	2	2	1		
	3–					1	2	5	1
	4–						2		1
	5–		1	1				1	
	6–								1
	7–				1	1			

Fig. 13. **Bivariate distribution.** The bivariate frequency table of the information compiled by a second-hand car dealer about a batch of 30 cars.

consequently, not make much sense to consider the distributions of the two variables separately; they are better looked at together as a bivariate distribution. The level of association between the two variables can be judged visually if a scatter diagram is drawn, or, in the case of linear association, calculated as a correlation coefficient.

The distributions obtained by adding along each row or column of a bivariate distribution table are called *marginal distributions*.

In the example of the second-hand cars (Fig. 13), the marginal distributions are:

Mileage (× 1,000)	0	10–	20–	30–	40–	50–	60–	70–
Frequency	1	2	2	3	7	6	6	3

and

Age	0–	1–	2–	3–	4–	5–	6–	7–
Frequency	3	2	7	9	3	3	1	2

If the probabilities of the two variables are considered, rather than their frequencies, the distribution is then a *bivariate probability distribution.*

If each row (or column) of a bivariate distribution table has the same variance, the distribution is said to be *homoscedastic* with respect to that variable, otherwise *heteroscedastic*. See BIVARIATE NORMAL DISTRIBUTION.

bivariate frequency table see CONTINGENCY TABLE.

bivariate Normal distribution a bivariate distribution in which each of the variables has a Normal distribution and where the correlation coefficient for the two distributions is known (it may be zero).

The bivariate Normal distribution has five parameters: the mean and variance of each of the separate variables, and the correlation coefficient. This is a very important bivariate distribution, both as a theoretical model and as a good representation of many practical situations, for example the distribution of lengths and weights of adult fish of a particular type.

When sample data from a bivariate Normal distribution are plotted on a scatter diagram, the resulting pattern is usually approximately elliptical (see Fig. 14).

The standard equation for the least squares regression line is usually only valid if the y-values are random but the x-values are non-random. However in the case where the x- and y-values are both random and are drawn from a bivariate Normal distribution, this equation is also valid.

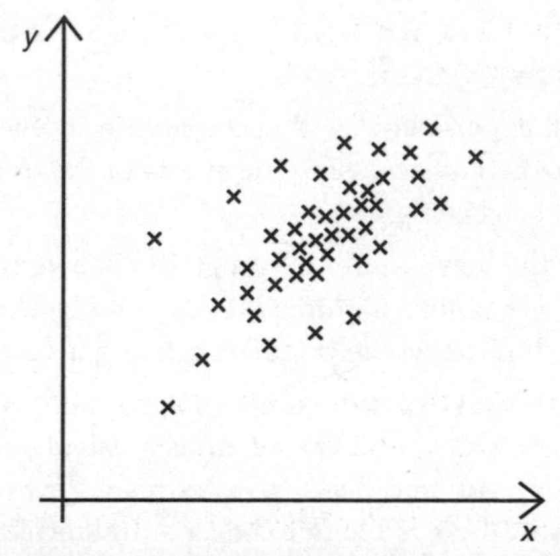

Fig. 14. **Bivariate Normal distribution.** The points form an approximately elliptical pattern.

The joint probability density function of this distribution is given by

$$f(x, y) = \frac{1}{2\pi\sigma_1\sigma_2\sqrt{1-\rho^2}} e^{-\frac{1}{2(1-\rho^2)}\left\{\frac{(x-\mu_1)^2}{\sigma_1^2} - 2\rho\frac{(x-\mu_1)(y-\mu_2)}{\sigma_1\sigma_2} + \frac{(y-\mu_2)^2}{\sigma_2^2}\right\}}$$

where $N(\mu_1, \sigma_1^2)$ and $N(\mu_2, \sigma_2^2)$ are the distribution of the two variables and ρ is their correlation coefficient. See NORMAL DISTRIBUTION, REGRESSION LINE.

blind experiment an experiment in which the subjects do not know whether they are being given the real treatment or a placebo (a dummy treatment).

In experiments on humans, there is a danger that the outcome may be affected by the perceptions of the people involved in the trial. For example, if a subject is told by a doctor that he or she is being given the latest and most expensive treatment, the psychological effect of this may be sufficient to bring about an improvement even if the treatment itself is ineffective . This is called the "placebo effect". To avoid this, a blind experiment is used in which the subjects are unaware of which treatment they are being given.

Another version of a blind experiment is used in situations where the measurement of the outcome of an experiment involves a subjective assessments by the subjects themselves. For example, in an experiment to assess the flavour of different brands of coffee, the tasters might be influenced by the reputation of the manufacturers of the coffee. Therefore, in this situation, it would be sensible to keep the names of the manufacturers hidden.

In a double blind experiment, neither the person receiving the treatment nor the person assessing the effects of the treatment knows which treatment has been given.

block a subset of the items under investigation in a statistical experiment. The variability within a block would be expected to be less than that within the whole population being investigated.

Example: the aim of an experiment is to compare the effects of several different fertilisers. An experimental field is available, divided into fairly small sections (often called *plots*). A stream flows alongside one edge of the field, so there is likely to be a natural fertility gradient sideways across the field; plots near the stream might well be

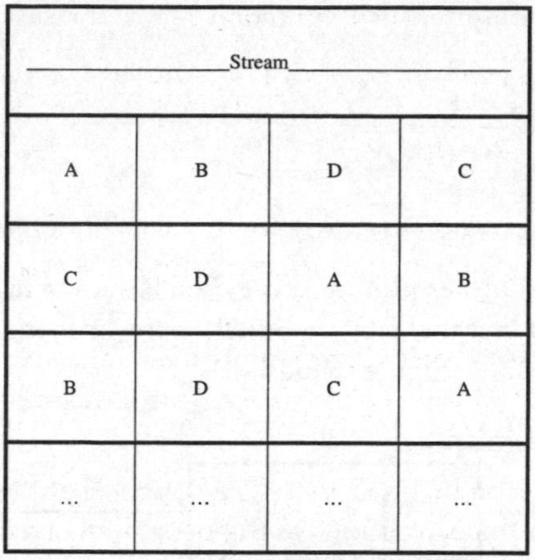

Stream			
A	B	D	C
C	D	A	B
B	D	C	A
...	...	...	...

Fig. 15. **Block.** A field divided into blocks.

naturally more fertile than those further away. So it is sensible to divide the field into strips running parallel to the stream, each strip having several plots in it, and to make sure that all the fertilisers are used in each strip, as shown in Fig. 15. The results within each strip can be compared with each other, but there may be consistent differences from strip to strip. In this experiment each strip is a block.

box and whisker plot, boxplot a diagram used to illustrate experimental data so as to bring out their important features. Box and whisker plots are often used in exploratory data analysis.

It consists of two lines (whiskers) drawn from points representing the extreme values to the lower and upper quartiles, and a box drawn between the quartiles which includes a line for the median.

Example: a train operating company collect data on the performance of a particular train on each of the 18 working days in a 3-week period. The data are minutes late, given to the nearest whole number.

	Monday	Tuesday	Wednesday	Thursday	Friday	Saturday
Week 1	3	4	4	15	1	4
Week 2	9	5	8	2	0	9
Week 3	11	1	6	3	7	2

For these data

Least value = 0
Lower quartile = 2
Median = 4
Upper quartile = 8
Greatest value = 15

The box and whisker plot for these data is shown in Fig. 16. It highlights the fact that the data are bunched towards the lower values but spread out to large higher values.

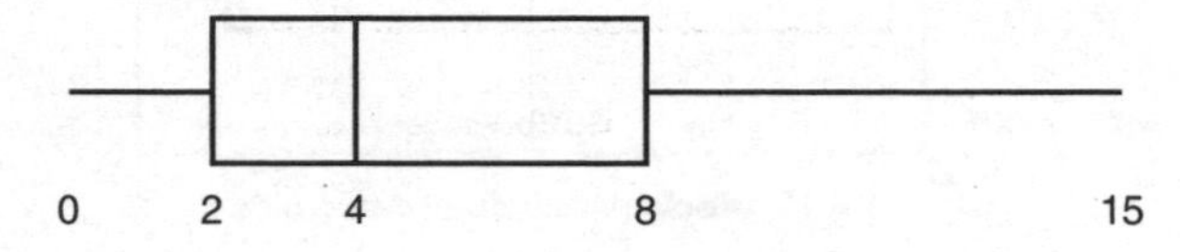

Fig. 16. **Box and whisker plot.**

When box and whisker plots are used in exploratory data analysis it is common practice to calculate the quartiles as *hinges* (see QUARTILE).

boxplot see BOX AND WHISKER PLOT.

Buffon's needle (of historical interest only) a statistically-based experiment for determining π. A piece of paper is marked with parallel lines, distance a apart, and a needle is thrown onto the paper in a random manner (see Fig. 17). The probability that the needle crosses one of the lines is given by:

$$\frac{2l}{a\pi}$$

where l is the length of the needle, and $l < a$. Thus if the needle is dropped a large number of times, it is possible to estimate π as

$$\frac{2l}{a} \times \frac{\text{Total number of throws}}{\text{Number of landings crossing a line}}$$

This experiment is a very slow way of finding π. After 10,000 throws, the first decimal should be known with reasonable confidence, the second not at all. It is also very difficult to ensure a random throw of the needle.

This experiment was originally intended as a method of calculating the stakes and winnings in the French gambling game franc-carreau, rather than a method of estimating π.

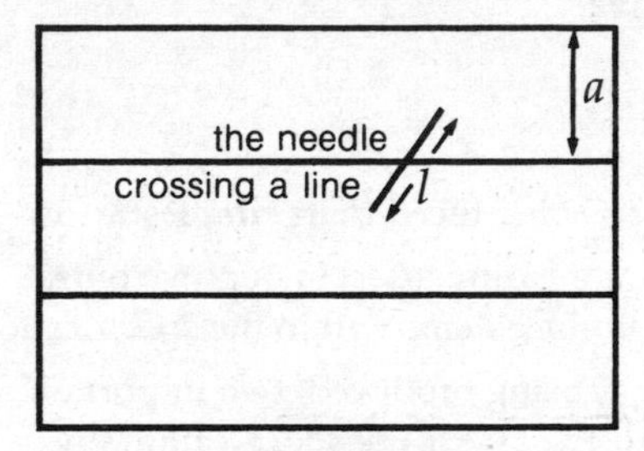

Fig. 17. **Buffon's needle.**

C

canonical having been reduced to its simplest form.

capability index a measure, used in quality control, of the capability of a machine to produce items within the specification.

When something is being produced, two important factors need to be matched up. The first is the specification given by the designer, and the second is the reality from the production line. For example, a designer might specify that the length of a component should be 27mm $\pm$1mm (these are the tolerance limits). If the machine being used can only produce items with lengths of 27mm $\pm$2mm there is a problem: either the designer needs to relax the tolerance limits, or a better machine needs to be found, or every item produced needs to be checked.

These two factors are brought together in the capability index, which is defined as

$$\textit{Capability index} = \frac{\textit{Upper tolerance limit} - \textit{Lower tolerance limit}}{6\sigma}$$

where σ is the standard deviation of the items being produced by the machine.

If the measure (length, weight, or whatever) being monitored is approximately Normally distributed, then almost all (99.7%) of the values will be within three standard deviations of the mean, μ. If the tolerance limits set by the designer happen to coincide with these values ($\mu \pm 3\sigma$), the process has a capability index of 1. This can be regarded as just satisfactory. If the mean of the production moves slightly, quite large numbers of defective items will be produced.

If, on the other hand, the tolerance limits set by the designer happen to coincide with the values $\mu \pm 6\sigma$, the process has a capability index of 2. This is very good because the mean of the production can move quite a lot without the items produced being defective.

capture-recapture a technique for estimating the number of animals (etc.) in a particular area.

Example: a zoologist wants to estimate the number of squirrels, N, on an island. She catches a sample of squirrels, of size n_1, marks them and then releases them so that they mix with the rest of the population. Some time later she catches a second sample, of size n_2, and counts the number, m, that are marked.

The proportion of marked squirrels in the second sample should be similar to the proportion of marked squirrels in the whole population.

So $\frac{m}{n_2} \approx \frac{n_1}{N}$, giving $N \approx \frac{n_1 n_2}{m}$.

The zoologist's first sample contains 50 squirrels. $n_1 = 50$

Her second sample (a week later) has 30 squirrels. $n_2 = 30$

In the second sample, 6 are marked. $m = 6$

She estimates the total number of squirrels on the island to be

$$N = \frac{n_1 n_2}{m} = \frac{50 \times 30}{6} = 250$$

Her answer of 250 squirrels is of course only a rough estimate.

The method assumes that the population is clearly defined and is not changing (rapidly) in size, and that the sample which has been marked mixes thoroughly with those which have not been marked.

cardinal number a number denoting quantity but not order in a group. Thus 1, 2, 3,... are cardinal numbers. The word cardinal is used in contrast to *ordinal* which expresses position in a sequence, e.g. first, second, third, etc. (see ORDINAL SCALE).

cartesian graph see GRAPH, ABSCISSA.

categorical data data referring to items which are in categories, for example the breeds of dogs in a show, rather than numerical, for example the ages of the dogs. See CATEGORICAL SCALE, NUMERICAL DATA.

categorical data analysis a collection of widely used and important methods for fitting models to categorical data.

categorical scale or **nominal scale** a scale which sorts variables according to category. Thus, the cars in a park can be sorted according to make, the fish in a pond according to type, etc. For a categorical scale it is essential that the categories be exhaustive, exclusive and clearly defined. Thus every variable should belong to one and only one category, and there should be no doubt as to which one.

There are two types of categories, a priori and constructed. *A priori categories* already exist, for example, a person's sex or year of birth; *constructed categories* are decided upon by research workers, for example a person's social class. A bad choice of categories can prejudice the outcome of an investigation. Categories may be indexed by numbers which do not correspond to any true order, like the numbers on footballers' shirts.

causation the production of an effect by a cause. Causation can never be proved statistically, although it may be very strongly suggested.

A common misunderstanding concerns the relationship between correlation and causation. If there is a strong correlation between the variables measuring the occurrence of two events A and B, it may be that

$$A \text{ causes } B,\ A \Rightarrow B$$
$$\text{or } B \text{ causes } A,\ B \Rightarrow A$$
$$\text{or they both cause each other, } A \Leftrightarrow B$$

However it may also be that a third event C causes both A and B,

$$C \begin{array}{l} \nearrow A \\ \searrow B \end{array}$$

or that a more complex set of interactions is going on.

Example: The fact that there is a strong (negative) correlation between the mean number of cars per family and the mean number of children per family for the various countries of the world does not mean that owning a car prevents you from having children, or that having children stops you having a car. The situation is more complicated since the affluence of the country is involved; in affluent countries, birth rates are low and car ownership is common.

$$\text{Affluence} \begin{array}{l} \nearrow \text{Low birth rate} \\ \searrow \text{High car owership} \end{array}$$

(It could, however, be argued that affluence and low birth rates both cause each other.)

The statement 'correlation does not imply causation' describes this situation.

cause variable see EXPLANATORY VARIABLE.

census **1.** a 100% survey.

2. an official periodic count of a human population, including such information as sex, age, occupation, etc. In the UK, a census is held every 10 years (2001, 2011, 2021,...). Heads of household are required to provide information about all those living under their roofs. The questions asked cover name, age, family status, education, and employment. A mini-census is sometimes taken in the interval between censuses using, typically, a 10% sample of the population.

Central Limit Theorem the theorem which states that, if samples of size n are taken from a parent population with mean μ and standard deviation σ, then the distribution of their means is approximately Normal, with

$$Mean = \mu, \quad Standard\ deviation = \frac{\sigma}{\sqrt{n}} \ (\text{or } Variance = \frac{\sigma^2}{n}).$$

As the sample size n increases, this distribution approaches the Normal distribution with increasing accuracy (see SAMPLING DISTRIBUTION). Thus, in the limit as n tends to infinity, the distribution of the sample means tends to

$$\mathrm{N}\left(\mu, \frac{\sigma^2}{n}\right).$$

Provided μ and σ are both finite, the Central Limit Theorem holds for samples from any parent population, whatever its distribution. In the special case where the parent population is itself Normal, the distribution of the sample means is Normal, whatever the sample size. (Notice that this special case is not the Central Limit Theorem.)

The Central Limit Theorem provides the basis for much of sampling theory.

certainty the condition of an event whose probability of occurrence is equal to 1.

chance **1.** the unknown and unpredictable element that causes an event to have one outcome rather than another.

2. that which is described by probability.

characteristic function see MOMENT GENERATING FUNCTION.

Charlier's check (of historical interest only) an accuracy check that can be used when calculating mean and standard deviation. It uses the fact that, for a set of numbers $x_1, x_2, \ldots x_n$, with frequencies $f_1, f_2, \ldots f_n$,

$$\sum_{i=1}^{n} f_i(x_i+1)^2 = \sum_{i=1}^{n} f_i x_i^2 + 2\sum_{i=1}^{n} f_i x_i + \sum_{i=1}^{n} f_i$$

and is illustrated in the following example.

Example: Find the standard deviation of

5, 5, 4, 4, 4, 3, 3, 2, 1, 1

x_i	f_i	$f_i x_i$	$f_i x_i^2$	Extra column $f_i(x_i+1)^2$
5	2	10	50	72
4	3	12	48	75
3	2	6	18	32
2	1	2	4	9
1	2	2	2	8
	$\sum f_i = 10$	$\sum f_i x_i = 32$	$\sum f_i x_i^2 = 122$	$\sum f_i(x_i+1)^2 = 196$

The extra column $f_i(x_i+1)^2$ is worked out and used to check $\sum f_i\,(=n)$, $\sum f_i x_i$ and $\sum f_i x_i^2$, all of which would normally be calculated anyway. In this case

$$\sum f_i(x_i+1)^2 = \sum f_i x_i^2 + 2\sum f_i x_i + \sum f_i$$
$$196 = 122 + 2 \times 32 + 10$$

which is easily seen to be true. It is then a simple matter to complete the calculations of mean and standard deviation, secure in the knowledge that the figures are correct.

chart see GRAPH.

Chebyshev's inequality the theorem which states that, if a probability distribution has mean μ and standard deviation σ, then the probability that the value of a random variable with that distribution differs from μ by more than $k\sigma$ is less than $\frac{1}{k^2}$, i.e.

$$P(|X-\mu| > k\sigma) < \frac{1}{k^2}$$

for any $k > 0$. The bound provided by Chebyshev's inequality is usually quite crude.

Chernoff faces a technique using cartoons or icons of the human face, with various features and expressions, to provide a visual display of multivariate data on up to 18 variables.

The "smiley face" symbol, ☺, is commonly used to mean "Good" and a down-turned mouth, ☹, to mean "Bad". This idea could be extended to represent more than two states; for example, 😐 could mean "Neither good nor bad" and other positions of the mouth could mean intermediate states.

This idea is taken much further in Chernoff faces by using not just the mouth but many other features of a face as well: the shape of the head, the size and shape of the eyes, the length of the nose and so on.

Much of the human brain is devoted to facial recognition and this gives people the ability to read the intended meaning in a Chernoff face.

chi-squared (or χ^2) **distribution** the distribution, with n degrees of freedom, of $\chi^2 = X_1^2 + X_2^2 + \ldots + X_n^2$, where $X_1, X_2, \ldots, X_n$ are n independent Normal random variables with mean 0 and variance 1.

The χ^2 distribution is used in a number of statistical tests including the chi-squared goodness of fit test, chi-squared test for variance, Friedman's two-way analysis of variance by rank and Kruskal-Wallis one-way analysis of variance.

Because the χ^2 distribution is different for each value of n (see Fig. 18) complete tables would be very cumbersome. They are therefore usually given only for critical values for those significance levels relevant to hypothesis tests. (e.g. 0.99, 0.95, 0.90, 0.10, 0.05, 0.02, 0.01; see Table 4).

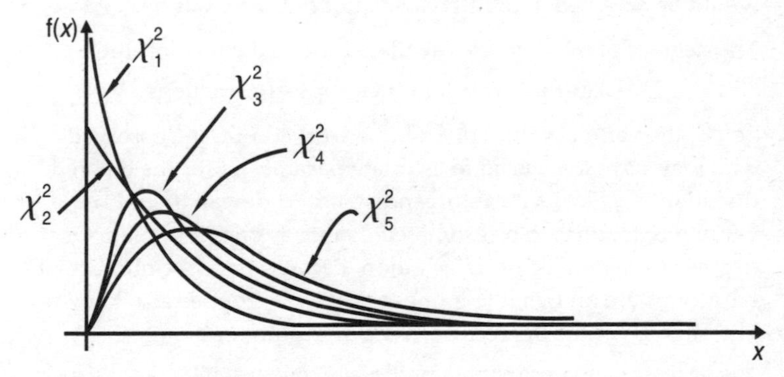

Fig. 18. χ^2 **distribution** for n = 1, 2, 3, 4 and 5.

chi-squared (or χ^2) **goodness of fit test** a hypothesis test of how well observed data and a given (theoretical) distribution are matched.

The null hypothesis is that the observed data are a sample drawn from a population with a given probability distribution.

The alternative hypothesis is that the data are drawn from a population which does not have that distribution.

The observed data are collected into groups, and the number of observations in each group denoted by f_o. The expected number in each group, f_e, is calculated from the given distribution. The test statistic X^2 is then worked out using

$$X^2 = \sum_{\substack{\text{all} \\ \text{groups}}} \frac{(f_o - f_e)^2}{f_e}$$

A small value of X^2 means the observed data fit the given distribution well, a large value badly.

The χ^2 test is a one-tail test. Critical values of X^2 for various significance levels are given in Table 4, for different degrees of freedom, ν, as in the extract below for $\nu = 5$.

Significance level, %	99.5	99	97.5	95	10	5	2.5	1	0.5	0.1
Critical value for $\nu = 5$	.412	.554	.831	1.15	9.24	11.07	12.83	15.09	16.75	20.52

Thus if, with 5 degrees of freedom, $X^2 = 10.4$, the null hypothesis would be accepted at the 10% level, but not the 5% level.

The degrees of freedom are calculated using the relationship

Degrees of freedom = Groups - Restrictions

Before the various values of f_e can be worked out, the observed data may have to be used to estimate parameters for the theoretical distributions, like its mean, μ, and standard deviation, σ. Each parameter that has to be estimated counts as one restriction. One degree of freedom is lost in addition, because once the numbers in k-1 groups are known, the number in the kth group is also known. The table gives the degrees of freedom in some common cases.

The χ^2 test is an approximate procedure, but usually a very good approximation unless any of the f_e values are very small. A rule of thumb that is very often used is that no value of f_e should be less than 5; if a value of f_e is less than 5, that group is combined with one of the other groups. However, it is also argued that lower values of f_e may be tolerated provided there are not too many of them.

Distribution	*Groups*	*Total sample size*	*Parameters needed*	*Total number of Restrictions*	*Degrees of freedom*
Normal	k	n	μ, σ	3	$k-3$
Binomial	k	n	p	2	$k-2$
Poisson	k	n	μ	2	$k-2$
Specified proportion	k	n		1	$k-1$
$r \times c$ contingency table	$r \times c$	n	Row and column totals	$(r-1)+(c-1)+1$	$(r-1)(c-1)$

Example 1: a die is thrown repeatedly with results as follows:

Score	1	2	3	4	5	6
Frequency	3	7	8	10	14	18

Is there evidence, at the 5% significance level, to suggest that the die is biased?

H_0: The die is unbiased; all six outcomes are equally likely.

H_1: The die is biased; the outcomes are not equally likely.

$$P(X = r) \neq \frac{1}{6} \text{ for } r = 1, 2, \ldots, 6$$

5% significance level

1-tail test

The total number of throws is

$$n = 3 + 7 + 8 + 10 + 14 + 18 = 60$$

So for each outcome, the expected frequency is $f_e = \frac{1}{6} \times 60 = 10$

Score	f_o	f_e	$f_o - f_e$	$\frac{(f_o - f_e)^2}{f_e}$
1	3	10	-7	4.9
2	7	10	-3	0.9
3	8	10	-2	0.4
4	10	10	0	0
5	14	10	4	1.6
6	18	10	8	6.4
			X^2	$= 14.2$

The only restriction is that the total $n = 60$. So the degrees of freedom are given by $\nu = 6 - 1 = 5$. (In this case, the expected distribution is in a specified proportion, 1:1:1:1:1:1.)

The critical value for X^2 for $v = 5$ at the 5% level (see Table 4) is 11.07.

Since 14.2 > 11.07, the null hypothesis is rejected. The evidence supports the suggestion that the die is biased.

Example 2: in a survey on class mobility, a research student interviews 200 men, asking them about their jobs and those of their fathers. He wishes to know whether the employment class of a man is related to that of his father. He classifies the results, using his own scale, as the following contingency table.

FATHER

SON

f_o	Upper	Middle	Lower	Total
Upper	16	25	19	60
Middle	15	33	22	70
Lower	19	22	29	70
Total	50	80	70	200

H_0: The employment class of a man is independent of that of his father.

H_1: The employment class of a man is not independent of that of his father.

10% significance level

1-tail test

According to the null hypothesis the expected numbers in the different categories would be as follows:

FATHER

SON

f_e	Upper	Middle	Lower	Total
Upper	$\frac{50}{200}\times 60$	$\frac{80}{200}\times 60$	$\frac{70}{200}\times 60$	60
Middle	$\frac{50}{200}\times 70$	$\frac{80}{200}\times 70$	$\frac{70}{200}\times 70$	70
Lower	$\frac{50}{200}\times 70$	$\frac{80}{200}\times 70$	$\frac{70}{200}\times 70$	70
Total	50	80	70	200

So the contingency table of expected frequencies is as follows:

FATHER

SON

f_e	Upper	Middle	Lower	Total
Upper	15	24	21	60
Middle	17.5	28	24.5	70
Lower	17.5	28	24.5	70
Total	50	80	70	200

The test statistic is calculated using $X^2 = \sum \frac{(f_o - f_e)^2}{f_e}$.

$$X^2 = \frac{1^2}{15} + \frac{1^2}{24} + \frac{(-2)^2}{21} + \frac{(-2.5)^2}{17.5} + \frac{5^2}{28} + \frac{(-2.5)^2}{24.5} + \frac{1.5^2}{17.5} + \frac{(-6)^2}{28} + \frac{4.5^2}{24.5}$$

$= 4.04$

The degrees of freedom for a 3×3 contingency table are

$v = (3 - 1) \times (3 - 1) = 4$.

The critical value for X^2 for $v = 4$ at the 10% significance level is 7.78.

Since $4.48 < 7.78$, there is no reason to reject the null hypothesis, that the employment class of a man is independent of that of his father.

Example 3: the scores of the football teams in a league one week are as follows.

Goals	0	1	2	3	4	5
Frequency	28	42	19	6	3	2

Carry out a test at the 5% significance as to whether these figures fit a Poisson distribution.

H_0: The underlying population is Poisson.

H_1: The underlying population is not Poisson.

5% significance level

1-tail test

The number of teams = 28 + 42 + 19 + 6 + 3 + 2 = 100

The number of goals $= 0 \times 28 + 1 \times 42 + 2 \times 19 + 3 \times 6 + 4 \times 3 + 5 \times 2 = 120$

Mean number of goals per team $= \frac{120}{100} = 1.2$

So if the null hypothesis is true, the Poisson parameter is given by $\lambda = 1.2$.

The expected frequencies, f_e, for 0, 1, 2, ... goals are thus given by the terms in

$$100e^{-1.2}\left(1+1.2+\frac{1.2^2}{2!}+\frac{1.2^3}{3!}+\frac{1.2^4}{4!}+\frac{1.2^5}{5!}+\ldots\right)$$

as given in the table below.

Goals	0	1	2	3	4	5	...
f_e	30.12	36.14	21.69	8.67	2.60	0.62	...

Since 2.60, 0.62 and subsequent expected frequencies are all less than 5, a single group for 3 goals or more is formed.

Goals	f_o	f_e	$\frac{(f_o-f_e)^2}{f_e}$
0	28	30.12	0.1492
1	42	36.14	0.9502
2	19	21.69	0.3336
≥3	11	12.05	0.0915
		Total X^2	1.5245

There are 4 groups and 2 restrictions (the total frequency and the mean), and so the degrees of freedom are given by $\nu = 4 - 2 = 2$.

The critical value for X^2 for $\nu = 2$ at the 5% significance level is 5.99.

Since 1.5245 < 5.99, there is no reason to reject the null hypothesis. The data are consistent with a Poisson distribution.

Data that fit too well.

The χ^2 tables (Table 4) also include figures at the other tail, 99%, 97.5%, 95% and 90%. These provide a different sort of check. It is sometimes argued that if the value of the test statistic X^2 is less than, say, the 95% value, the distribution fits the data under test so well as to arouse suspicion. The situation should be examined critically to see if one of three things has happened:

(a) the data have been invented;

(b) the theoretical distribution has been formed round the data, and then the data are being used to confirm the same theory;

(c) data which do not fit well have been rejected.

If none of these is the case, then the data and the distribution are indeed well matched.

chi-squared (or χ^2) **test for variance** a test of the null hypothesis, that a particular sample has been drawn from a population with a given variance. This test strictly speaking requires the parent population to be Normal, but it is often a good approximate procedure if this is not the case. It depends upon the fact that the statistic

$$\frac{(n-1)s^2}{\sigma^2}$$

has a χ^2 distribution with $n-1$ degrees of freedom, where n is the sample size, s is the sample standard deviation (calculated with divisor $n-1$) and σ is the given population standard deviation.

Example: The sample 9, 5, 6, 4, 3, 5, 4, 6, 2 is drawn from a Normal population. Test, at the 5% significance level, whether it could have been drawn from a population with variance 2.

For this test

H_0: $\sigma^2 = 2$

H_1: $\sigma^2 \neq 2$

5% significance level

2-tail test

From the sample data,

$$\bar{x} = \frac{44}{9} \text{ and } s^2 = \sum_{i=1}^{n} \frac{(x_i - \bar{x})^2}{n-1} = 4.11$$

$$\text{So } \frac{(n-1)s^2}{\sigma^2} = \frac{8 \times 4.11}{2} = 16.44$$

The degrees of freedom are $\nu = 9 - 1 = 8$.

The critical values of χ^2 are found from Table 4.
For 8 degrees of freedom they are 2.18 (the 97.5% value) and 17.53 (the 2.5% value). (See Fig. 19.)

Since 16.44 lies between 2.18 and 17.53, the null hypothesis is accepted at this level.

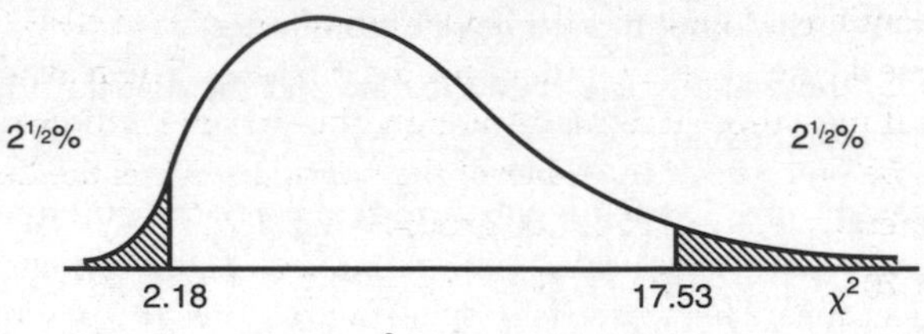

Fig. 19. χ^2 **test for variance.**

Confidence limits for variance: The χ^2 distribution can also be used to work out confidence limits for the population variance. As with χ^2 test for variance, the underlying population should be Normal distribution but the procedure often provides a good approximation when this is not the case. In the example, the sample

9, 5, 6, 4, 3, 5, 4, 6, 2

was drawn from a Normal population. The population variance was estimated as:

$$s^2 = 4.11$$

95% confidence limits for an estimate are given by:

$$\frac{(n-1)s^2}{x_1} \text{ and } \frac{(n-1)s^2}{x_2}$$

where x_1 (= 2.18) and x_2 (= 17.53) are the relevant critical values for 8 degrees of freedom (see Fig. 13).

Thus the 95% confidence limits are:

$$\frac{8 \times 4.11}{2.18} \text{ and } \frac{8 \times 4.11}{17.53}$$

15.09 and 1.88

When the variances of two samples are used to test if those samples could have been drawn from populations with the same variance, the *F* test is used. See CONFIDENCE INTERVAL.

class a collection or division of people, things, or values of a variable, sharing a common characteristic. See GROUPED DATA. See also CLASS BOUNDARIES.

class boundaries the boundaries for the classes when data are grouped.

For a continuous variable, the upper boundary of one class may be the same as the lower boundary for the next one. The reason for this is that, if measured accurately enough, the variable will virtually always be one side or the other of the boundary. Thus the heights of adult men (in metres) could be grouped as:

1.65 - 1.70 1.70 - 1.75 1.75 - 1.80 etc.

In practice, however, measurements are always rounded, in some way or other, and so it is often convenient to make it clear which class a variable apparently on the boundary is to be assigned to. So it is usual to use one of the following conventions for describing class intervals.

1.65 - 1.70 - 1.75 -

$1.65 \le h < 1.70$ $1.70 \le h < 1.75$ $1.55 \le h < 1.80$

or the equivalent conventions for when the upper boundary is included, namely

- 1.70 - 1.75 - 1.80

$1.65 < h \le 1.70$ $1.70 < h \le 1.75$ $1.55 < h \le 1.80$

For a discrete variable, on the other hand, class boundaries cannot be shared. The distribution of the annual egg yield per hen of a farmer's chickens might, for example, be grouped as:

No. of eggs	0-49	50-99	100-149	150-199	200-249	250-299	300-349	≥350
Frequency (no. of hens)	1	15	35	46	53	42	18	0

In this case the same figures do not appear in the boundaries of two classes. There is no need for this, since the number of eggs laid must be a whole number.

classification placing objects into classes, groups or categories using defined criteria.

class interval the interval between class boundaries for grouped data, or the length of this interval

climograph a graph showing rainfall against temperature for a particular place. Points are plotted for each month of the year (mean values being taken), and then joined up to form a continuous 12-sided figure (see Fig. 20).

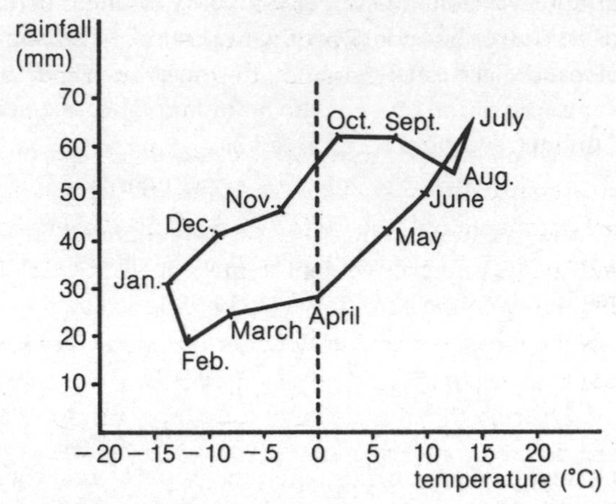

Fig. 20. **Climograph.** Mean rainfall and temperature for Arkhangelsk.

closed question a question with only a limited number of possible responses. By contrast, an *open question* has any number of possible responses.

On a questionnaire, closed questions are usually accompanied by an answer box to be ticked or filled in, as in the following examples.

- What is your age in completed years? ☐
- Are you male or female? M/F (delete as appropriate)
- How did you find the speaker?

☐ Very Good ☐ Quite good ☐ Fair ☐ Poor ☐ Very poor

By contrast, it is usual to allow respondents some space to write their responses to open questions.

- What do you think about the performance of the current government?

When designing a questionnaire, you have to decide whether to use closed questions or open questions, or a combination of the two.

When analysing a questionnaire, closed questions are much easier to deal with than open questions. Open questions can be very informative, but the responses need to be classified in some way if they are to be analysed.

Open questions can often be transformed into closed questions by deciding in advance the most likely responses (possibly following a pilot survey) and including the option "Other". However, respondents may feel frustrated if they are not able to give the response which they feel really represents their position.

cluster analysis a collection of statistical techniques used in multivariate analysis to assign items to groups or clusters, with the intention of reducing the dimension of the data. Clustering is done on the basis of a defined measure of the "distance" or "similarity" between the items with close or similar items being placed in the same cluster. The effect is often to produce a *hierarchy* of clusters. The process can be illustrated by a *dendrogram*.

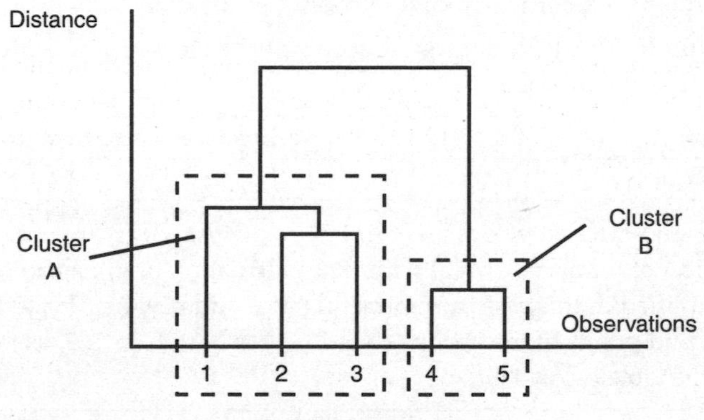

Fig. 21. **Dendrogram.**

cluster sampling a method of sampling in which the members of the sample are chosen from one or several groups (or clusters) rather than at random from the entire population. From a statistical point of view this is less satisfactory than simple random sampling, but it may well be more practical and / or economical.

Imagine a veterinary research officer investigating the incidence of various parasites among goats in Nigeria. To select, say, 100 random goats in the country, and travel backwards and forwards to see each one, would be very time-consuming and expensive; indeed, in the absence of some form of register of goats, it would be impossible to achieve truly random selection. What he could do instead would be to select several villages at random from a register of villages, and then select some goats at random at each village. This would be cluster sampling; the villages would be the clusters. He might even take all 100 from the same village, but that might be unsatisfactory, as it could, for instance, be the case that all the goats in that village had infected each other with the same parasites.

coded data data which have been translated from the form in which they were collected to one that is more convenient for analysis. For example, in questionnaires, people are often asked to reply by ticking appropriate boxes. When the replies are analysed, each box usually has a number (its code) associated with it, and it is these numbers that are analysed rather than the ticks themselves.

When figures are grouped, it is often possible to simplify calculations on them by a particular kind of coding of the data.

Example: This table shows the scores of the members of a cricket team during a season.

Score (runs)	0-9	10-19	20-29	30-39	40-49	50-59	60-69	70-79
Frequency	21	24	18	13	12	6	4	2

The mid-interval values are $4\frac{1}{2}$, $14\frac{1}{2}$, … $74\frac{1}{2}$, which are not the easiest of numbers to deal with. The calculation of the mean can be simplified in this case by subtracting $4\frac{1}{2}$, and dividing by 10 to give the mid points of the groups coded values of 0, 1, 2, … 7, as is shown in the table below below.

Coded value	0	1	2	3	4	5	6	7
Frequency	21	24	18	13	12	6	4	2

For the coded values, the mean is $\frac{215}{100} = 2.15$ and this value is uncoded to give the mean of the data: $2.15 \times 10 + 4.5 = 26$.

The calculation of standard deviation may similarly be simplified by the use of coding.

The coding could also have been done by subtracting, say $34\frac{1}{2}$ from each mid-interval point, and then dividing by 10, which would have given coded values -3, -2, -1, 0, 1, 2, 3 and 4. This would have reduced the size of the numbers in the calculation still further.

In general, for $y = a + bx$, $\bar{y} = a + b\bar{x}$ and $s_y^2 = b^2 s_x^2$.

coefficient **1.** an index of measurement of a characteristic. Examples are the COEFFICIENT OF ALIENATION, and the COEFFICIENT OF DETERMINATION.

2. a numerical or constant factor in an algebraic expression. For example, 5 and 7 are the coefficients of x^2 and x^3 respectively in the expression $5x^2 + 7x^3$.

coefficient of alienation a measure of the extent to which two random variable are unrelated. The coefficient of alienation is given by

$$\sqrt{(1-r^2)}$$

where r is Pearson's product-moment correlation coefficent.

coefficient of determination the proportion of the variation of the dependent variable which is taken up by fitting the REGRESSION LINE (or other regression model). Notation r^2, sometimes R^2. The variation in the set $y_1, y_2, \ldots, y_n$, with mean $\bar{y}$, is given by

$$\sum_{r=1}^{n} (y_r - \bar{y})^2.$$

This can be split into two parts:

$$\sum_{r=1}^{n} (y_r - \bar{y})^2 = \underbrace{\sum_{r=1}^{n} (y_r - \hat{y}_r)^2}_{\text{Unexplained Variation}} + \underbrace{\sum_{r=1}^{n} (\hat{y}_r - \bar{y})^2}_{\text{Explained Variation}}$$

where $(x_r, \hat{y}_r)$ is the point on the y on x regression line vertically above or below the point (x_r, y_r). Thus $\hat{y}_r$ is the predicted value of y_r in the linear regression of y on x. Thus the unexplained variation is the sum of the squares of the residuals $\varepsilon_1, \varepsilon_2, \ldots \varepsilon_n$.

The coefficient of determination is thus defined as

$$\textit{Coefficient of determination} = \frac{\text{Explained variation}}{\text{Total variation}}$$

Example: find the coefficient of determination for the least squares regression line for points (1,3), (3,5), (5,6), (7,10), shown in Fig. 22.

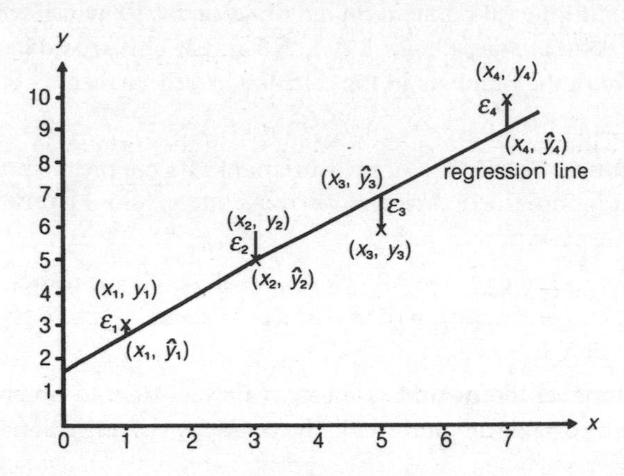

Fig. 22. **Coefficient of determination.** Graph showing the residuals ε_1, ε_2, ε_3 and ε_4 of the set of points $(x_1,y_1) \ldots (x_4,y_4)$ for the regression line.

x	y	$x-\bar{x}$	$y-\bar{y}$	$(x-\bar{x})^2$	$(y-\bar{y})^2$	$(x-\bar{x})(y-\bar{y})$
1	3	-3	-3	9	9	9
3	5	-1	-1	1	1	1
5	6	1	0	1	0	0
7	10	3	4	9	16	12
16	24			$S_{xx}=20$	$S_{yy}=26$	$S_{xy}=22$

$\bar{x}=4$ $\bar{y}=6$

From these figures, the regression line can be calculated:

$$y-\bar{y}=\frac{S_{xy}}{S_{xx}}(x-\bar{x}) \qquad y-6=\frac{22}{20}(x-4)$$

which simplifies to $y = 1.1x + 1.6$. The calculation then continues as follows:

x	y	$\hat{y}=1.1x+1.6$	$d=(y-\hat{y})$	d^2
1	3	2.7	0.3	0.09
3	5	4.9	0.1	0.01
5	6	7.1	-1.1	1.21
7	10	9.3	0.7	0.49
16	24			1.8

Total variation $\sum(y-\bar{y})^2 = 26.0$

Unexplained variation $\sum d^2 = \underline{1.8}$

Explained variation $= 24.2$

$$\textit{Coefficient of determination} = \frac{\text{Explained variation}}{\text{Total variation}} = \frac{24.2}{26}$$
$$= 0.931$$

The calculation can be simplified by using the formula

$$\textit{Coefficient of determination} = \frac{S_{xy}^2}{S_{xx}S_{yy}}.$$

In this example $\frac{S_{xy}^2}{S_{xx}S_{yy}} = \frac{22^2}{20 \times 26} = 0.931.$

The coefficient of determination has the same value as r^2, where r is the number that would result from a calculation of the product moment correlation coefficient. However the regression line is usually used in cases when one variable (y) is random and the other (x) is non-random whereas the CORRELATION COEFFICIENT applies when both variables are random. Thus, in a context in which it is appropriate to use the coefficient of determination, it is not right to describe r as the correlation coefficient, even though it is calculated using the same formula. The coefficient of determination may, however, be thought of in terms of the correlation coefficient between the observed and fitted values.

coefficient of kurtosis see KURTOSIS.

coefficient of mean deviation see ABSOLUTE MEASURE OF DISPERSION.

coefficient of skewness see SKEW.

coefficient of variation see ABSOLUTE MEASURE OF DISPERSION.

combinations the ways of selecting a subset of a set, where the order in which the elements are selected is of no importance. Thus the possible combinations of three letters out of A, B, C, D and E are 10 in all:

A B C	A B D	A B E
A C D	A C E	A D E
B C D	B C E	B D E
C D E		

The number of ways of selecting r objects from a total of n, all different, is denoted by:

$$\binom{n}{r} = \frac{n!}{r!(n-r)!}$$

An alternative notation for is $\binom{n}{r}$ is ${}^nC_r$.

The quantity $\binom{n}{r}$ also gives the coefficient of $x^r\, y^{n-r}$ and $x^{n-r}\, y^r$ in the binomial expansion of $(x + y)^n$. See also FACTORIAL, BINOMINAL COEFFICIENTS.

complementary events events which are both exclusive and exhaustive so that one of them must happen, but only one. The events, 'The snake is alive' and 'The snake is dead' are complementary. They cannot both happen at the same time (they are exclusive), and they cover all possibilities (they are exhaustive).

complementary sets see SETS.

completely randomised design see EXPERIMENTAL DESIGN.

concordance see KENDALL'S COEFFICIENT OF CONCORDANCE.

conditional probability the probability p of an event occurring, given that another event has occurred. The usual notation for the combined event of 'Event A occurring given that event B has occurred' is $A|B$ (A given B). The probability is given by:

$$P(A|B)=\frac{P(A\cap B)}{P(B)}$$

Example: a card player cheats by looking at one of his opponent's cards. He sees a black picture card (King, Queen, or Jack) but he cannot make out which one. What is the probability that the card is a King?

Event A: The card is a King.

Event B: The card is a black picture card.

$$P(A) = \frac{4}{52} \qquad P(B) = \frac{6}{52} \qquad P(A\cap B)=\frac{2}{52}$$

$$P(A|B)=\frac{P(A\cap B)}{P(B)}=\frac{\frac{2}{52}}{\frac{6}{52}}=\frac{1}{3}$$

Conditional probability can be used in the description of independent events. Events A and B are independent if $P(A|B)=P(A|B')$ and $P(B|A)=P(B|A')$, where A' is the event not -A and B' is not -B.

conditioning event an event which is known to have occurred. The conditional probability of event A occurring given that event B has occurred is given by:

$$P(A|B)=\frac{P(A\cap B)}{P(B)}$$

In this, B is the conditioning event.

confidence interval an interval within which a parameter of a population is estimated, on the basis of sample data, to lie. Usually the larger the sample size n, the smaller is the confidence interval; in other words, the more accurate is the estimate of the parameter.

Example: a biologist traps 20 frogs of a particular species and weighs them. These are the measurements in grams.

22, 26, 30, 34, 47, 48, 50, 50, 52, 53
53, 53, 55, 57, 57, 61, 64, 72, 74, 82

The mean of these figures is 52, and so it is estimated that the mean mass of a frog is 52 g, i.e. that 52 g is the mean of the underlying population. It is however, very unlikely that the true population mean is exactly 52 g, although it will probably be somewhere near that figure. The question which therefore arises is, 'How accurate is the figure of 52 g?'

The 95% confidence interval is the interval produced in such a way that if a large number of samples of 20 frogs are taken, and a 95% confidence interval for the population mean is worked out from each sample, then 95% of the confidence intervals will include the true population mean; the other 5% will not.

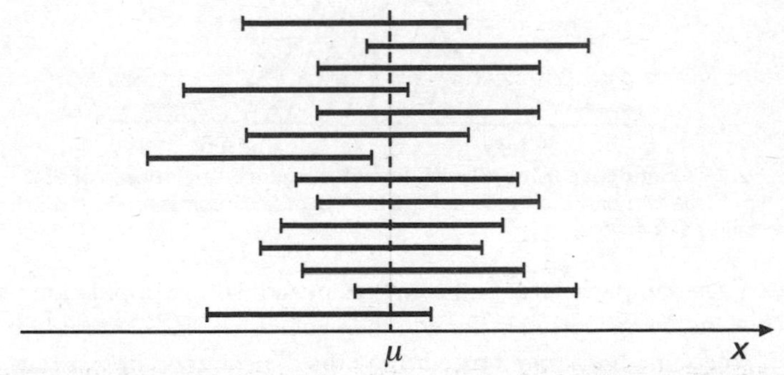

Fig. 23. **Confidence interval.** The diagram shows the confidence intervals obtained from different samples; most, but not all, of them capture the true population mean μ.

This idea is illustrated in Fig. 23. A number of samples of a given size have been collected and each one has been used to construct a confidence interval for the population mean, μ. Most of the intervals do in fact contain μ but one does not.

The upper and lower bounds of the 95% confidence interval are the 95% confidence limits for the population mean. Confidence limits may be taken for other levels, e.g., 90%, 99%, 99.9%, etc.

Confidence limits for the mean of a distribution are worked out using the formula:

$$\bar{x} \pm k \frac{\text{"Population standard deviation"}}{\sqrt{n}}$$

where k is a number whose value depends on the confidence level, and the circumstances of the sampling.

- If the sample is from a Normal population whose standard deviation is known, k is found from normal distribution tables (Table 2). Commonly used values for k are:

  $k = 1.645$ for 90% limits
  $k = 1.96$ for 95% limits
  $k = 2.58$ for 99% limits

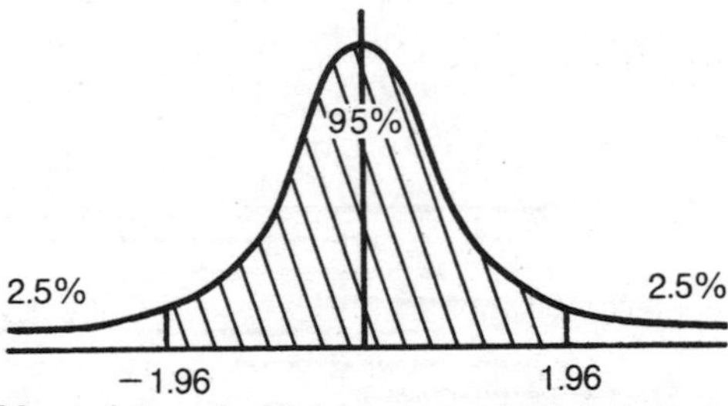

Fig. 24. **Confidence interval with known standard deviation.** For 95% limits, the two unshaded areas in Fig. 24 are both 2.5%, corresponding to ± 1.96 standard deviations.

- If the sample is large and comes from some other population whose standard deviation is known, then because of the Central Limit Theorem, the above procedure will still be a good approximation.

In cases where the population standard deviation is unknown, it has to be estimated from the sample data using:

$$s^2 = \frac{1}{n-1} S_{xx} \text{ where } S_{xx} = \sum_{i=1}^{n} (x_i - \overline{x})^2 = \sum_{i=1}^{n} x_i^2 - n\overline{x}^2$$

- If the sample is large, then this estimate is likely to be quite accurate. Then, again using the Central Limit Theorem, the above procedure will still be a good approximation.
- If the sample is small, then if it is from a Normal population the values of k are found from the t distribution with $n - 1$ degrees of freedom.
- If the sample is small, and it is not from a Normal population, using the t distribution is at best an approximation and might be a dangerously poor one. No simple procedure of this kind is available in such cases.

There is no firm rule as to how large a "large sample" has to be. In many circumstances 30 is likely to be large enough. Smaller sample sizes are sometimes adequate.

In the example of the frogs, the sample size of 20 is not really very large, nor is the population standard deviation known. So the t distribution is used, with 20-1=19 degrees of freedom. Using this requires the assumption that the underlying population is Normal.

For $\nu = 19$, and 95% limits, the table of critical values of t (see Table 3) gives $k = 2.093$.

The population standard deviation is estimated using:

$$s = \sqrt{\frac{\sum_{i=1}^{n} (x_i - \overline{x})^2}{n-1}} = 15.40$$

So the 95% confidence limits for the population mean are:

$$52 \pm 2.093 \times \frac{15.40}{\sqrt{20}}, \text{ giving } 44.8 \text{ and } 59.2$$

The 95% confidence interval for the parent mean is 44.8 to 59.2.

This is the interval within which it is "reasonable" to suppose the true value of the mean mass of the frog population to lie.

Sample data can also be used to work out confidence intervals for other parameters of the parent population, for example the variance.

Binomial sampling experiments are used to estimate p, the probability of success in each trial. Confidence limits for p may be calculated directly or found by using a suitable confidence interval chart.

Example: a binomial experiment is conducted 50 times, with 32 successes. The probability of success p is estimated as:

$$\hat{p} = \frac{32}{50} = 0.64$$

95% confidence intervals for p are found from a chart for binomial sampling experiments (Fig. 25) by locating $\hat{p} = 0.64$ on the horizontal axis; finding the intersections of the vertical line through 0.64 with the two curves marked 50 (the number of trials); then reading the equivalent values of p (0.49 and 0.77) on the vertical scale. Thus, in this example, the 95% confidence limits are 0.49 and 0.77.

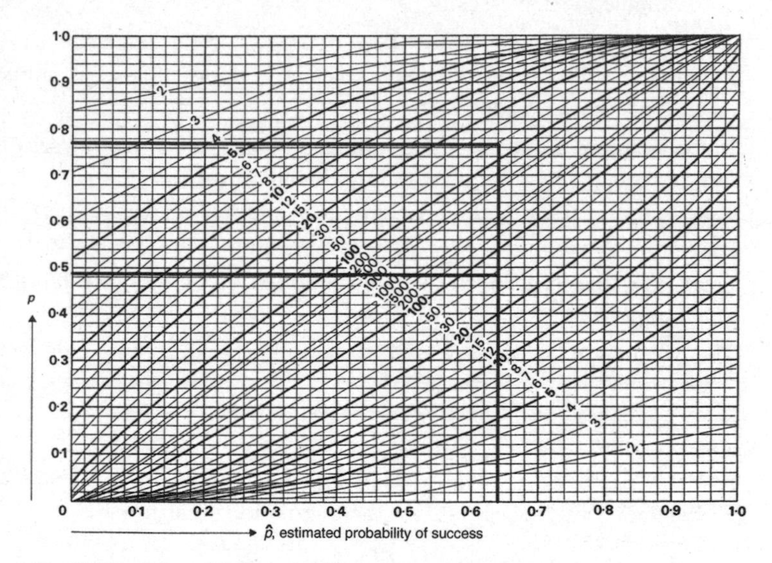

Fig. 25. **Confidence interval.** 95% confidence interval chart for binominal sampling experiments. Source: *Elementary Statistics Tables*, H.R. Neave, George Allan and Unwin.

For the use of confidence interval charts for population correlation coefficients, see FISHER'S z TRANSFORMATION.

confidence limits the lower and upper boundaries of a CONFIDENCE INTERVAL.

conformance criteria see ACCEPTANCE SAMPLING.

confounding describing a situation that arises in factorial experiments, and elsewhere, when it is not possible to distinguish a potential effect of interest from some other nuisance effect that was not allowed for when designing the experiment.

Example: the purpose of an experiment is to compare the effects of two fertilisers, *A* and *B*, on the yield of a crop. Several plots of land are marked out in a field: *A* is applied to some, *B* to the others. After completing the experiment, it is realised that all the *A* plots are near a stream flowing along one side of the field, and all the *B* plots are further away. In the analysis of the data, it is impossible to tell whether any apparent differences between the yields of the *A* plots and the *B* plots is because one of *A* and *B* is actually better or because of a natural fertility gradient in the soil of the field (caused by the stream). The two effects are *confounded*.

consistent (of an estimator) such that the larger the sample size, the more precise is the estimator. An estimator $\hat{\theta}_n$ of the value, θ, of a population parameter is formed on the basis of a sample of size n. The estimator is said to be consistent if, as n increases, the limit of $\hat{\theta}_n$ is the true value, θ.

constant a quantity, or its algebraic representation, whose value does not vary.

consumer's risk customer's risk (in quality control) the probability of a consignment with the proportion of defectives equal to the lot tolerance percentage defective being accepted, despite the inspection scheme. This is thus the probability of a borderline batch being allowed through. Clearly, the larger the sample size (meaning more thorough sampling), the lower the consumer's risk should be. Historical note: consumer's risk was at one time a name for Type 2 error.

contingency table or **bivariate frequency table** a table showing a bivariate frequency distribution. The classifications may be quantitative or qualitative. See also BIVARIATE DISTRIBUTION.

Example: The distribution of candidates entering a particular examination, given by age and by sex.

		AGE OF CANDIDATE							
		< 15	15	16	17	18	19	20	≥ 21
SEX OF CANDIDATE	*Female*	100	6,322	13,449	9,387	2,969	1,045	592	2,812
	Male	164	6,077	12,212	7,909	2,779	1,285	824	3,176

continuity corrections corrections that may need to be applied when the distribution of a discrete random variable is approximated by that of a continuous random variable.

Example: Intelligence Quotient, as measured by certain tests, is an integer. It is thus a discrete variable. Its distribution has mean 100 and standard deviation 15.

Its distribution may, however, be approximated by the Normal distribution, with the same mean and standard deviation. This in effect involves fitting the Normal distribution curve to a frequency

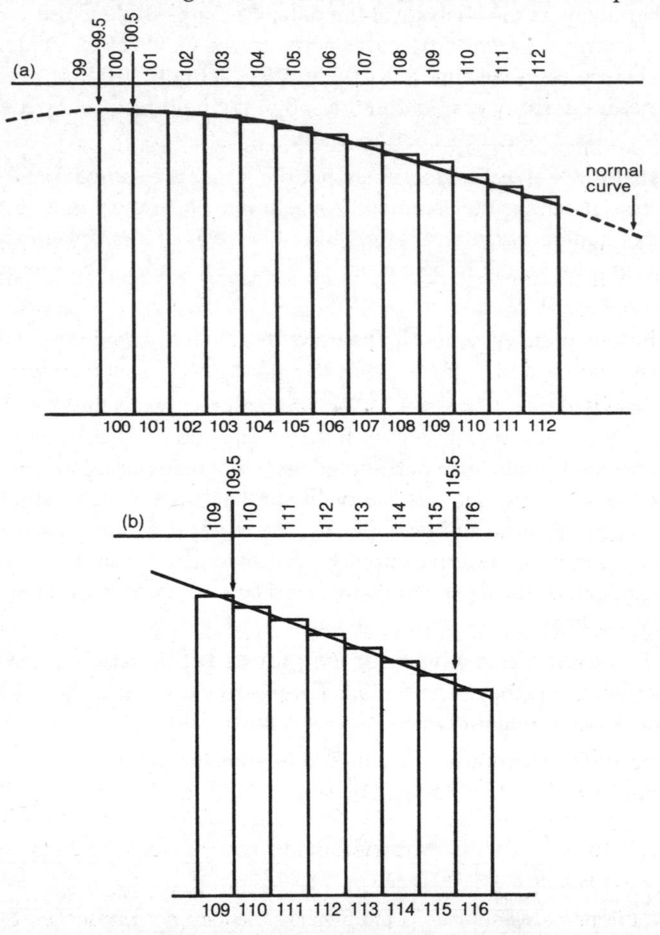

Fig. 26. **Continuity corrections.** (a) and (b).

chart (Fig. 26(a)). When this is done, the space that meant 100 on the frequency chart takes on the meaning 99.5-100.5 for the Normal curve.

The question, 'What proportion of people have an IQ between 110 and 115 (inclusive)?', could be answered by finding the sum of the heights of the bars labelled 110 to 115. If, however, the Normal distribution is used, the correct interval is 109.5 to 115.5. This correction of 0.5 at either end of the required interval is called a continuity correction (Fig. 26(b)). The area is then:

$$\Phi\left(\frac{115.5-100}{15}\right)-\Phi\left(\frac{109.5-100}{15}\right)=0.113$$

Thus, 11.3% of people have an IQ between 110 and 115.

When the binomial distribution is approximated by the Normal distribution, a continuity correction may be required. Yates' correction is a continuity correction sometimes applied in the χ^2 test for 2×2 contingency tables to allow for the continuity of the χ^2 distribution.

continuous variable a variable which may take any value (between reasonable limits); this contrasts with a discrete variable, which may take only certain values. Thus the height of an adult woman is a continuous variable, whereas the number of goals scored by a hockey team is a discrete variable. It is possible for a woman to have any reasonable height, like 1.68 m, but a hockey team cannot score 2.3 goals.

Some variables, while technically discrete, may best be treated as continuous because the steps are so small. The weekly pay of British men is discrete with steps of 1 p, the smallest unit of currency, but this is so small compared to the figures involved that, for most practical purposes, such a variable is continuous.

The distinction between continuous and discrete random variables may be seen in terms of probabilities. The probability of a continuous variable taking any particular value *exactly* is zero. The question, 'What is the probability that a woman (selected at random) is exactly 1.68 m tall?', is meaningless, having answer zero. A more sensible question would be, 'What is probability that a woman (selected at random) is between 1.675 and 1.685 m tall?'. By contrast, it is quite sensible to ask 'What is the probability that a die shows exactly 5 when it is thrown?'

Distinctions between these variables are, however, complicated by the fact that it is possible for a variable to have a mixed distribution with both continuous and discrete components, as in the following example.

Example: a bus leaves a stop every 10 minutes, having waited there 2 minutes. The distribution of times waited by passengers before boarding the bus can be discrete (time = 0, they arrive when the bus is there) or continuous (times up to 8 minutes; they have to wait for the next bus). The variable, the time waited, is however, continuous. If the distribution of a continuous variable is not continuous, then the probability of it taking a particular exact value need not be zero, as in the case of time 0 when waiting for the bus.

control charts (in quality control) diagrams used to give warning of when a machine or process is not functioning well. In this context, manufacturing processes may be divided into two types. There are those whose product is either good or faulty, like electric light bulbs, and others where the quality of the product is determined by a measurement. If a machine for making 5 cm nails is producing them with a mean length of 4.8 cm, such nails are unsatisfactory, as they

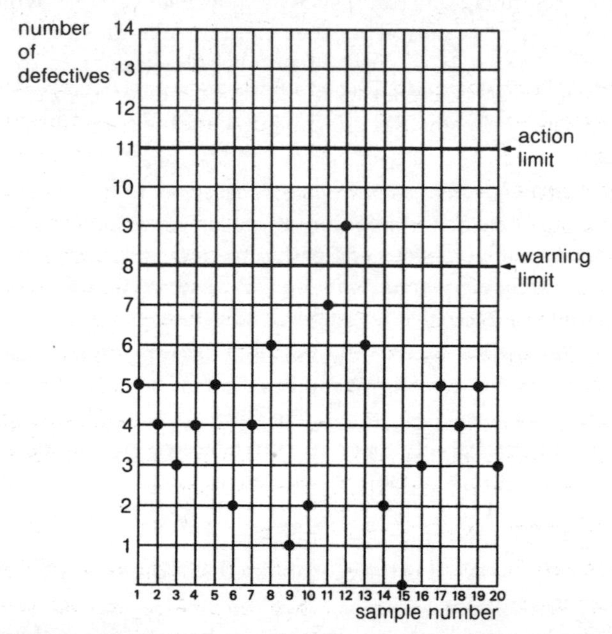

Fig. 27. **Control charts.** The chart for numbers of defectives in samples of size 100 (process average, 4%).

do not meet the specification; they are, however, still usable as nails, whereas faulty light bulbs are quite worthless.

In the situation where the product is either good or faulty, an inspection scheme would involve recording the numbers of defectives in samples of a certain size. This information can then be plotted on a control chart. This chart typically has two lines drawn on it, the inner control (or warning) limit and the outer control (or action) limit; see Fig. 27. The warning limit is usually set so that if the process is running correctly it is exceeded by 5% of the samples, the action limit by 0.1%.

In cases where a measurement is involved, two things need to be watched, the mean of the quantity and its variability. A change in either could result in an unduly high proportion of the products being outside acceptable tolerance limits. Two control charts are thus needed, one for the mean, the other for the spread (range or standard deviation) of the samples. The control limits are typically set so that 5% of items from a correctly functioning machine lie outside the warning limits. If the standard deviation is not known it is often estimated using Table 10, for converting range into standard deviation.

Control charts may also at times be used for measurements from individual items, rather than for the means and variability of the samples.

control experiment an experiment designed to check or correct the results of another experiment by removing one of the variables operating in that other experiment. If, for example, the effects of a sleeping pill were being tested, one group of subjects would be observed after being given the pill. In the control experiment, another equivalent group, the *control group*, would be observed after not being given it.

In that example it would be good practice to give the control group a pill that should have no effect, a placebo, but for the subjects to be kept ignorant of which pill they had been given.

control group see CONTROL EXPERIMENT.

correlation linear association between two random variables.

If two variables are such that, when one changes, the other does so in a related manner, they are said to be associated. If the association is linear, the two variables are correlated.

The level of correlation in a bivariate population with variables X and Y is given by the standardised population covariance and denoted by ρ. The equivalent measure for sample data is Pearson's product-moment correlation coefficient, r.

In everyday speech, the term correlation is sometimes applied to non-linear association but this is technically incorrect. Non-linear association can be measured using a rank correlation coefficient, either Spearman's or Kendall's, but the formulae used for these measures are actually based on linear association between the ranks.

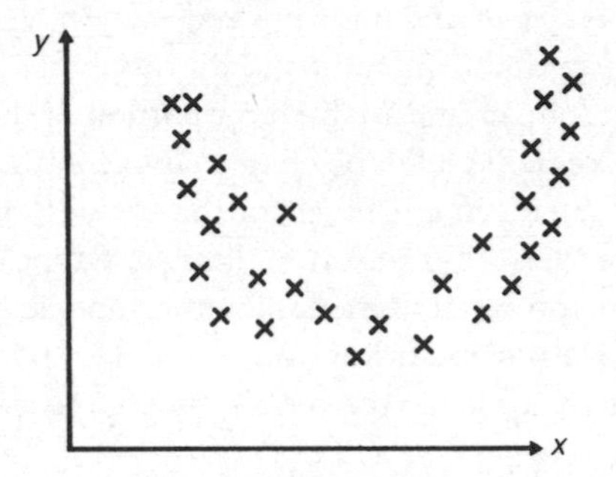

Fig. 28. **Correlation.** Non-linear association is not correctly described as correlation.

correlation coefficient a measure of linear association between two random variables, or their ranks. A correlation coefficient is a number between -1 and 1. A coefficient of 1 means perfect positive correlation, -1 perfect negative correlation, and 0 no correlation.

Population correlation. The level of correlation between random variables X and Y is given by the standardised population covariance divided by the product of the population standard deviations:

$$\rho = \frac{\mathrm{Cov}(X,Y)}{\sigma_X \sigma_Y}$$

Sample correlation: Pearson's product moment-correlation coefficient

The equivalent measure for sample data is Pearson's product-moment correlation coefficient, r, and this is calculated using the formula

$$r = \frac{S_{xy}}{\sqrt{S_{xx} S_{yy}}}$$

where $S_{xx} = \sum_{i=1}^{n}(x_i - \bar{x})^2$, $S_{yy} = \sum_{i=1}^{n}(y_i - \bar{y})^2$, and $S_{xy} = \sum_{i=1}^{n}(x_i - \bar{x})(y_i - \bar{y})$.

This is a measure of linear association and so is unsuitable if the relationship is non-linear. Both variables must be random.

Example: a fruit grower wishes to know if there is any correlation between the rainfall in the month of March and the ultimate yield. He keeps records of rainfall in this month and the average yield of his trees for five years.

	2003	2004	2005	2006	2007
x (Rainfall in cm)	28	32	29	41	30
y (Yield in kg)	134	142	136	168	150

The calculation is as follows:

	x	y	$x-\bar{x}$	$y-\bar{y}$	$(x-\bar{x})^2$	$(y-\bar{y})^2$	$(x-\bar{x})(y-\bar{y})$
	28	134	−4	−12	16	144	48
	32	142	0	−4	0	16	0
	29	136	−3	−10	9	100	30
	41	168	9	22	81	484	198
	30	150	−2	4	4	16	-8
$\div n$	5) 160	5) 730			$S_{xx} = 110$	$S_{yy} = 760$	$S_{xy} = 268$
	$\bar{x} = 32$	$\bar{y} = 146$					

Thus the correlation coefficient r is:

$$r = \frac{268}{\sqrt{110\ x\ 760}} = .927$$

Rank correlation coefficients: Spearman's coefficient of rank correlation

There are occasions when it is not convenient, economic, or even possible, to give values to variables, only rank orders. In such cases, a rank correlation coefficient has to be used. *Rank correlation coefficients* may also be appropriate in cases of non-linear relationship between the variables.

Spearman's coefficient of rank correlation is given the symbol r_S. It is calculated by the formula:

$$r_s = 1 - \frac{6\sum d^2}{n(n^2 - 1)}$$

where d is the rank difference for any member.

Example: in a dog show there are two judges, O'Hara and O'Shaughnessy. They place the dogs as follows:

	Annie	Buster	Cleo	Dolly	Empress
O'Hara	1	5	2	4	3
O'Shaughnessy	2	4	1	3	5
d	-1	1	1	1	-2
d^2	1	1	1	1	4

The value of d is obtained simply by subtracting O'Shaughnessy's rank from that of O'Hara, for each dog.

Thus $\sum d^2 = 1+1+1+1+4=8$ and $n=5$, giving Spearman's coefficient of rank correlation as

$$r_s = 1 - \frac{6\sum d^2}{n(n^2 - 1)} = 1 - \frac{6 \times 8}{5(25 - 1)} = 0.6$$

Kendall's coefficient of rank correlation is given the symbol τ. It is calculated for n pairs of values using the formula:

$$\tau = \frac{S}{\frac{1}{2}n(n-1)}$$

where S is a sum of scores; its computation is explained in the example below.

Example: taking the same data for the dog show used in the previous example, the calculation of Kendall's coefficient of rank correlation is as follows:

	Annie	Buster	Cleo	Dolly	Empress
O'Hara	1	5	2	4	3
O'Shaughnessy	2	4	1	3	5

One of the sets of results is put in order (in this case that from O'Hara), and the other is placed beneath it.

	Annie	Cleo	Empress	Dolly	Buster
O'Hara	1	2	3	4	5
O'Shaughnessy	2	1	5	3	4

Each entry in the second row (i.e. from O'Shaughnessy) is compared in turn with each of the entries to the right of it. If the entry to the right is greater +1 is counted, if it is less -1, and if it is equal, 0.

Thus the entry of 2 under Annie scores a total of 2 when it is compared with the entries for the four other dogs.

Cleo	Empress	Dolly	Buster		
-1	+1	+1	+1	=	2
(1 < 2)	(5 > 2)	(3 > 2)	(4 > 2)		

The total score, *S*, is the sum of the scores for each dog, in this case,

S	=	2	+	3	+	-2	+	1	=	4
		Annie		Cleo		Empress		Dolly		

The last entry, Buster, having no numbers to the right of it, makes no contribution to the total score.

The calculation is then completed using the formula

Kendall's rank correlation coefficient, $T = \dfrac{S}{\frac{1}{2}n(n-1)}$.

In this case $S = 4$ and $n = 5$, giving $T = 0.4$.

The bottom line of the formula, i.e. $\frac{1}{2}n(n-1)$, represents the maximum possible score which would occur if there were perfect agreement (in this case 10).

Interpretation of correlation coefficients.

Sometimes a correlation coefficient is worked out because its value is of interest, at other times as a test statistic for a hypothesis test on the correlation coefficient for the underlying bivariate population from which the sample was drawn.

Test of possible correlation.

The usual null hypothesis for this test is that the correlation coefficient of the underlying population is zero. Thus the points taken are a sample from a bivariate population in which the two variables are uncorrelated.

Example: using the earlier example of the fruit trees,

Rainfall (cm)	28	32	29	41	30
Yield (kg)	134	142	136	168	150

For this test

$H_0 : \rho = 0$
$H_1 : \rho \neq 0$

10% significance level
2-tail test

There are 5 pairs of results, and the (Pearson's product-moment) correlation coefficient $r = 0.927$.

The critical value is found from tables. Some care is needed here because some tables are given for n, the sample size and others for ν, the degrees of freedom; $\nu = n - 2$. Table 6 is given for n, and so the critical value for a 2-tail test at the 10% significance level is 0.805

n	.10	.05	.02	.01
5	.805	.878	.934	.959

$0.927 > 0.805$ and so H_0 is rejected. The evidence does not support the hypothesis of no correlation between rainfall and yield.

The use of this test involves the assumption that the underlying population is bivariate Normal.

The three different correlation coefficients given in this entry, r, r_s and τ, all have their own tables of significance (see Tables 6, 7 and 8).

For the use of a confidence interval chart for r, see Fig. 50 (FISHER'S z TRANSFORMATION).

Test for non-zero correlation coefficient.

To test the null hypothesis that the underlying population has a product moment correlation coefficient other than zero, Fisher's z transformation is used. This also allows confidence limits to be set up for an estimated value of the population correlation coefficient. See FISHER'S z TRANSFORMATION.

correlogram see AUTOCORRELATION.

covariance a measure of the linear association between two random variables. The term covariance can denote a parameter of a bivariate population or a statistic derived from a bivariate sample.

Population Covariance

The population covariance of two random variables, X and Y, is defined as

$$\text{Cov}(X, Y) = \text{E}\big[(X - \text{E}(X)\ \big(Y - \text{E}(Y)\big)\big]$$

Alternatively this can be written as

$$\text{Cov}(X, Y) = \text{E}(XY) - \mu_X \mu_Y.$$

The covariance of independent random variables is zero. Consequently, the relationship for variance of the sum or difference of two random variables, X and Y,

$$\text{Var}(X \pm Y) = \text{Var}(X) + \text{Var}(Y) \pm 2\text{Cov}(X, Y)$$

reduces to

$$\mathrm{Var}(X \pm Y) = \mathrm{Var}(X) + \mathrm{Var}(Y)$$

if X and Y are independent.

Sample Covariance

For a sample $(x_1, y_1), (x_2, y_2),...(x_n, y_n)$, the sample covariance is given by:

$$\mathrm{cov}(x, y) = \frac{S_{xy}}{n} = \frac{1}{n}\sum_{i=1}^{n}(x_i - \bar{x})(y_i - \bar{y})$$

For a sample from a bivariate distribution in which the variables are independent, the expected value of the sample covariance is zero.

According to how they are written, the formulae for the equation of a regression line and the product-moment correlation coefficient can involve the sample covariance. However, it is common in modern usage to write these formulae in terms of the sums of squares and the sum of products, S_{xx}, S_{yy} and S_{xy} without explicit involvement of covariance.

Cramér-Rao lower bound an expression for the smallest possible variance for an unbiased estimator.

critical region the set of values of a test statistic which leads to the rejection of the null hypothesis in a hypothesis test. The form taken by the critical region depends on the particular test, whether it is 1-tail or 2-tail, the significance level and, in some cases, the degrees of freedom.

As an example, in a test on the mean of a Normal distribution, the test statistic is z. In the case of a 2-tail test with a 5% significance level, the critical region is $z < -1.96$ or $z > 1.96$. These are the shaded regions in Fig. 29.

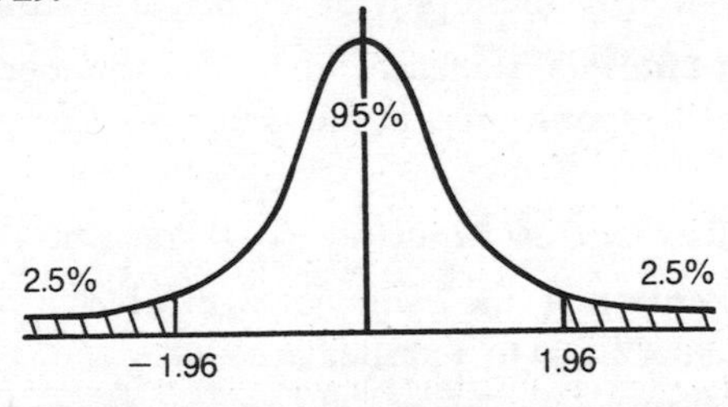

Fig. 29. **Critical region.** For a 2-tail text using the Normal distribution at the 5% significance level, the tails of 2.5% represent values more extreme than ±1.96 standard deviations from the mean.

The values of the test statistic which determine whether or not the null hypothesis is accepted are called the critical values; in the example above the critical values are –1.96 and 1.96.

critical value see CRITICAL REGION.

cross-over design an experimental design involving two treatments in which the experimental material is the same for both treatments.

When comparing two treatments, the experimental material to which they are applied should be as alike as possible so that any difference in the results is because of the treatments and not because of differences in the experimental material. In a cross-over design the experimental material is the same for both treatments, and so there should be no variation.

Example: two treatments, *A* and *B*, for lowering blood pressure are being compared. In a cross-over design, half of the patients (chosen randomly) are given treatment *A* then treatment *B*, and the other half are given treatment *B* then treatment *A*. This gives a paired design and the two treatments are compared on each of the patients.

The potential problem with this design is that there may be carry-over effects. Maybe treatment *A* has a long lasting effect. If this is used first, treatment *B* will appear to have little effect because the blood pressure is already low. To try to allow for this possibility there is usually a waiting time before the second treatment is given. It is also possible to look for this in the analysis of the results.

With a group of 20 patients available, the cross-over design gives 20 pairs of results. The alternative is to divide the people randomly into two groups and give one group treatment *A* and the other group treatment *B*, but this would provide a much less sensitive test.

cross-section analysis statistical analysis of variations in data obtained from a set of sources at the same time. Compare TIME SERIES ANALYSIS.

cumulative distribution function see DISTRIBUTION FUNCTION.

cumulative frequency the frequency with which a variable has value less than or equal to a particular value.

Example: the yield, y kg, of each tree in an orchard of 200 trees is recorded, and the figures are grouped as follows:

Yield, y (kg)	0<*y*≤10	10<*y*≤20	20<*y*≤30	30<*y*≤40	40<*y*≤50	50<*y*≤60	60<*y*≤70	70<*y*≤80	80<*y*≤90
Frequency (trees)	0	4	16	32	46	48	34	12	8

The corresponding cumulative frequency table is:

Yield	*y*≤10	*y*≤20	*y*≤30	*y*≤40	4*y*≤50	*y*≤60	*y*≤70	*y*≤80	*y*≤90
Frequency	0	4	20	52	98	146	180	192	200

This is derived from the frequency table by adding the frequencies from the left, or cumulating up to the entry in question. This can be drawn as a cumulative frequency graph or *ogive* (Fig. 30). Such a graph allows the median, quartiles, percentiles, etc., to be read easily. It also allows questions like 'Estimate how many trees yielded less than 43 kg,' to be answered readily. If the points are joined with straight lines rather than a curve, the diagram is called a *cumulative frequency polygon*.

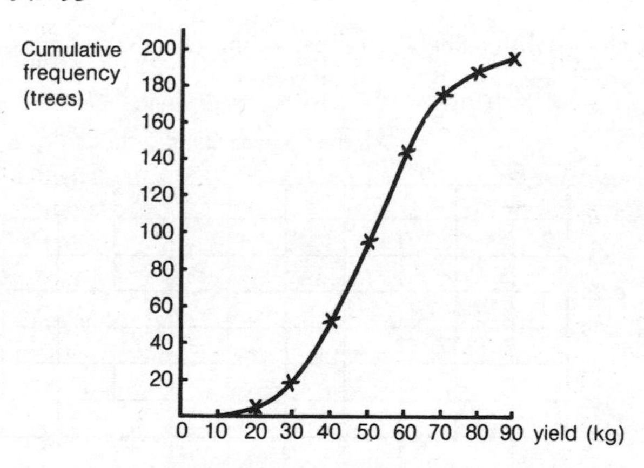

Fig. 30. **Cumulative frequency graph, or ogive.**

See TABLE 2(a), DISTRIBUTION FUNCTION.

customer's risk see CONSUMER'S RISK.

CuSum charts charts used in quality control to monitor a process.

CuSum is an abbreviation for "cumulative sum". CuSum charts are used in monitoring a process to see whether it is on target.

Example: the diameter of a particular component should be 50 mm. Samples are taken from the production line at regular intervals and the mean diameter of the components in the sample is calculated. The deviations from the 50 mm target are calculated and the cumulative sum of these deviations is plotted as a time series, as in Fig. 31.

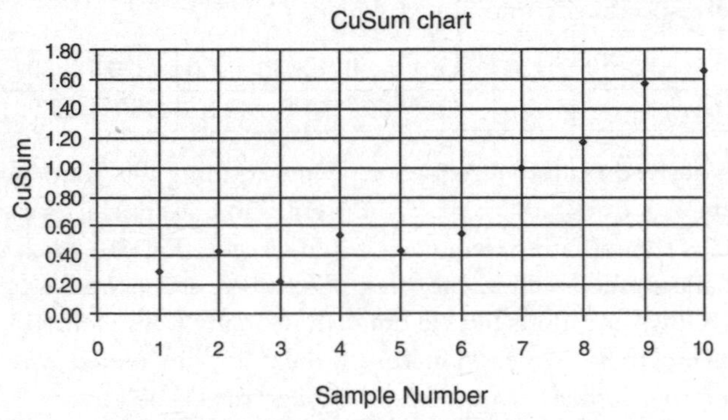

Fig. 31. **CuSum chart.** Cumulative deviations for 10 samples.

This contrasts with a Shewhart means chart (Fig. 32) where successive values are plotted as a simple time series.

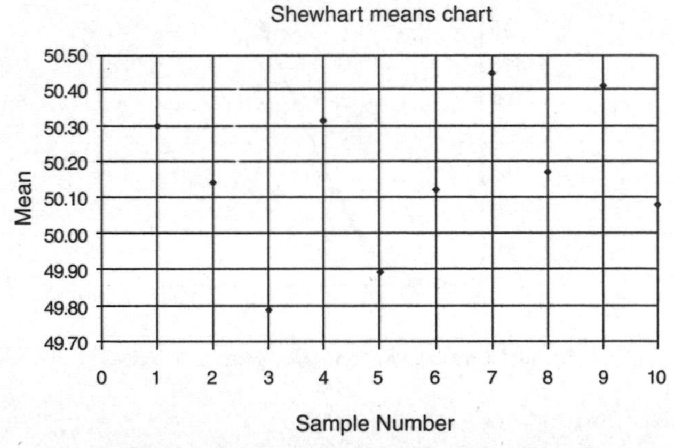

Fig. 32. **Shewhart means chart.** Means for 10 samples.

CuSum charts are used in the same situations as Shewhart means charts, but they are more sensitive to the mean being off target. However, they are more difficult to interpret on-line, as it is the gradient which is important, not the actual value. Because of this, they are often used retrospectively to identify the time at which a process went off target. The hope is that it will be possible to find out what happened at that time, an assignable cause.

cyclic component a non-seasonal component in a time series which varies in a recognizable cycle. An example is the frequency of sunspots, which has an 11-year cyclic component (Fig. 33).

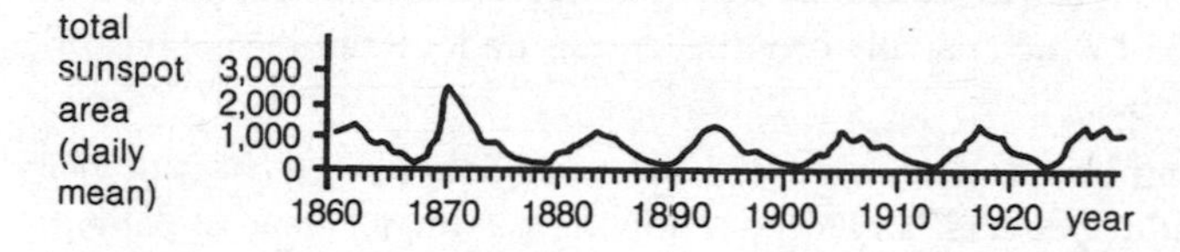

Fig. 33. **Cyclic component.** The 11-year cycle of solar activity (after W.J.S. Lockyer).

See TIME SERIES.

d

data information obtained from a survey, an experiment or an investigation.

database a store of information, often held on a computer. Information in a database is stored as a number of records or files, each of which usually contains entries under a number of headings or fields.

Example: a doctors' surgery keeps, as a data base, records of visits from its patients, under the following headings: Name of patient; Name of doctor; Sex; Date of birth; Date of visit (day, month, year); Symptoms; Diagnosis; Treatment.

The ability of a computer to cross-reference within a data base makes it a particularly efficient form for storing and handling data. It could, for example, be made to search the database for all patients born between 1960 and 1970 diagnosed as suffering from varicose veins.

data capture **1.** any process for converting information into a format that can be handled by a computer.

2. a process of gathering data.

datum zero a base level from which values of a variable are taken. For example, heights of tides are given with respect to a datum zero, called Low Water Ordinary Springs (LWOS). For many years it was defined physically by a mark on the harbour wall at Newlyn in Cornwall, England. Dates are measured with respect to a datum zero, the start of year 1 AD. Longitude is measured with respect to the Greenwich meridian as datum zero.

In the assumed mean method of working out averages, the assumed mean is a datum zero for the numbers involved. See ASSUMED MEAN.

death rate the ratio of deaths in a specified area, group, etc., to the population of that area, group, etc., usually expressed per thousand per year.

decile one of nine actual or notional values of a variable dividing its distribution into 10 groups with equal frequencies.

decimal places the figures to the right of the decimal point that are required to express the magnitude of a number to a specified degree of accuracy.

Example: 3.141 592 7 written to four decimal places is 3.1416; i.e., there are four digits after the decimal point, namely 1, 4, 1, and 6. In this case the final digit, 6, has been rounded up from 5 because the following digit, 9, is at least 5.

decision rule see ACCEPTANCE SAMPLING.

degrees of freedom the number of free variables in a system, often denoted by v. There are many situations where correct knowledge of the number of degrees of freedom is essential. These often involve the use of tables at the end of a calculation or procedure.

The idea of degrees of freedom can be seen from this obviously ridiculous argument.

"Tim and Roy are two boys. Their heights are 164 cm and 156 cm respectively, and in their end-of-term mathematics test they score 80% and 40%. Looking at these figures, someone concludes that 'Tall boys are better at mathematics than short ones'. He illustrates his theory on a graph (Fig. 34(a)). He then works out the correlation coefficient, and finds it to be 1, providing (he claims) "the final confirmation of his theory."

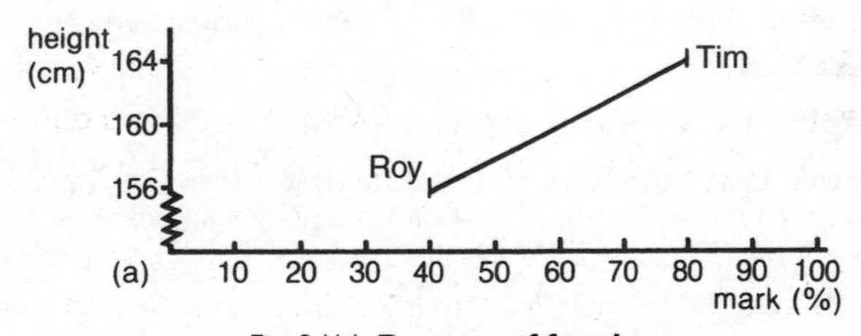

Fig. 34(a). **Degrees of freedom.**

The flaw in his argument is that if he takes only two points, unless they happen to be equal, he will always get perfect correlation (positive or negative) because it takes two points to define his theory. Only when he takes a third pair of values is he in a position to carry out a test at all. To continue the example, suppose a third boy, Harry was 162 cm tall and scored 30% on the test.

When his point is added to the graph (Fig. 34(b)) it is not found to lie near the line defined by the other two. Although there are three points on the graph, two are used in defining the line (or theory); only one is actually being used in testing it. There are thus 3 – 2 = 1 degrees of freedom in the system.

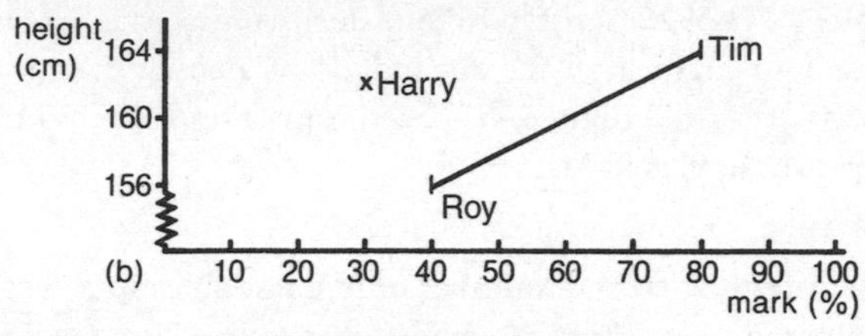

Fig. 34(b). **Degrees of freedom.**

In correlation, the number of degrees of freedom is similarly given by $v = n - 2$, where n is the number of points. It is not, however, usually possible to pick out two particular lost points. It is more appropriate to think of the set as a whole having lost 2 from its total of n independent variables.

As a general rule, for n observations for a model

Degrees of freedom = n – number of restrictions

where a restriction occurs for each parameter of the model that has to be estimated from the observed data.

The degrees of freedom for the various hypothesis tests covered in this dictionary are given under the relevant entries. See CHI SQUARED TEST, MODELLING.

demography the science of population statistics.

de Morgan's laws the laws which state that, for events or sets, A_1, $A_2, \ldots A_n$,

$$\text{(a)} \quad \left(\bigcup_{i=1}^{n} A_i\right)' = \bigcap_{i=1}^{n} A_i'$$

$$\text{(b)} \quad \left(\bigcap_{i=1}^{n} A_i\right)' = \bigcup_{i=1}^{n} A_i'$$

These may be stated in words as:

1st Law 'The complement of the union of events is the same as the intersection of their complements'

2nd Law 'The complement of the intersection of events is the same as the union of their complements.'

These laws are illustrated by this simple example from sets.

$$\mathcal{E} = \{a, b, c, d, e, f\}, A_1 = \{a, b, c\} \text{ and } A_2 = \{b, c, d\}.$$

de Morgan's 1st Law

$A_1 \cup A_2 = \{a, b, c, d\}$, and so $(A_1, \cup A_2)' = \{e, f\}$;

$A_1' = \{d, e, f\}$, $A_2' = \{a, e, f\}$, and so $A_1' \cap A_2' = \{e, f\}$;

so $(A_1 \cup A_2)' = A_1' \cap A_2'$.

de Morgan's 2nd Law

$A_1 \cap A_2 = \{b, c\}$, and so $(A_1 \cap A_2)' = \{a, d, e, f\}$

$A_1' \cup A_2' = \{a, d, e, f\}$

So $(A_1 \cap A_2)' = A_1' \cup A_2'$.

dendrogram see CLUSTER ANALYSIS.

density estimation a method for estimating a probability density function directly from data. Several techniques are available, usually based on some notion of smoothing a histogram.

dependent variable, response variable a variable which depends on another variable. The variable upon which it depends is called the *independent*, or explanatory variable. When a graph is drawn, the dependent variable is usually drawn on the vertical (y) axis, the independent variable on the horizontal (x) axis (Fig. 34). The term dependent variable may also be used in cases where it depends on more than one variable.

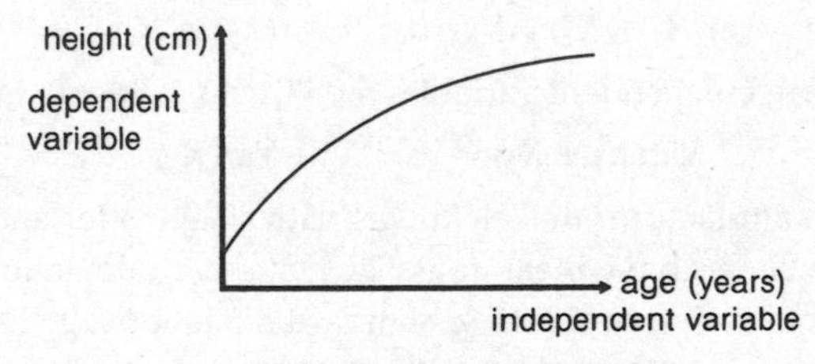

Fig. 35. **Dependent variable.**

Fig. 35 illustrates the graph of a girl's height, measured every birthday, against her age. Her height is the dependent variable, her age the independent. Her height depends on her age but her age does not depend on her height; it is just a statement of the time she has been alive.

descriptive statistics that branch of statistics which involves describing, displaying or arranging data. Pie charts, bar charts, pictograms, etc., are all used in descriptive statistics. Compare with INFERENTIAL STATISTICS and INTERPRETIVE STATISTICS.

deterministic model a model in which all events are the inevitable consequences of antecedent causes.

deviation the difference between a value of a variable and the mean of its distribution.

Example: the numbers 3, 6, 7, 10 and 4 have mean 6, and so their deviations are:

$$3 - 6 = -3$$
$$6 - 6 = 0$$
$$7 - 6 = 1$$
$$10 - 6 = 4$$
$$4 - 6 = -2$$

The sum of all deviations in a set is zero. Deviation is used in the calculation of variance, standard deviation, covariance, etc., but is not often used in its own right.

difference and sum of two or more random variables the random variables which result from subtracting or adding two or more random variables. The mean and variance of the random variables $X_1 \pm X_2$, formed from the two random variables, X_1 and X_2, are given by the following relationships:

Mean: $E(X_1 \pm X_2) = E(X_1) \pm E(X_2)$

Variance: $\mathrm{Var}(X_1 \pm X_2) = \mathrm{Var}(X_1) + \mathrm{Var}(X_2) \pm 2\mathrm{Cov}(X_1, X_2)$.

If X_1 and X_2 are independent variables then $\mathrm{Cov}(X_1, X_2) = 0$. In that case

$$\mathrm{Var}(X_1 \pm X_2) = \mathrm{Var}(X_1) + \mathrm{Var}(X_2)$$

Example: a manufacturer makes knives with steel blades and wooden handles. The blades have mean mass 20 g, standard deviation 1.5 g; the handles have mean mass 15 g, standard deviation 2 g. The blades and handles to be assembled are chosen independently. What are the mean, variance, and standard deviation of the mass of a knife?

Mean = 20 + 15 = 35 (g)

Variance = $1.5^2 + 2^2 = 6.25$ (g^2)

Standard deviation = $\sqrt{\text{Variance}} = \sqrt{1.5^2 + 2^2} = 2.5$ (g)

Example: a joiner buys pieces of timber with mean length 300 cm, standard deviation 7.5 cm. He then cuts off pieces with mean length 250 cm, standard deviation 4 cm. What are the mean and standard deviation of the lengths of the offcuts?

$$\textit{Mean} = 300 - 250 = 50\text{cm}$$
$$\textit{Standard deviation} = \sqrt{7.5^2 + 4^2} = 8.5 \text{ cm}$$

These rules can be extended to more than two variables.

$$\text{E}(X_1 \pm X_2 \pm ... \pm X_n) = \text{E}(X_1) \pm \text{E}(X_2) \pm \ldots \pm \text{E}(X_n)$$
$$\text{Var}(X_1 \pm X_2 \pm ... \pm X_n)$$
$$= \text{Var}(X_1) + \text{Var}(X_2) + \ldots + \text{Var}(X_n) + \sum_{\text{all pairs } r,s} \pm 2\text{Cov}(X_r, X_s).$$

dimension the number of variables in a multivariate distribution.

discrete consisting of distinct or separate parts.

discrete variable a variable which may take only certain discrete values; compare CONTINUOUS VARIABLE. The number of people resident in a household is a discrete variable which may have value 1, 2, 3, etc., but not intermediate values, such as 1.5 or 2.446. On the other hand, the weight of the oldest member of the household is a continuous variable, since it can take any reasonable value and is not restricted in this way.

discriminant analysis a technique used in multivariate analysis for distinguishing between two or more defined groups. The technique works by using a combination of the existing variables to create a new variable that maximises the differences between the groups. The technique can then be used to assign unknown cases to their appropriate groups.

disjoint see SET.

dispersion the spread of a distribution. In the diagram (Fig. 36) distribution *B* has a greater dispersion or spread than *A* although they both have the same mean.

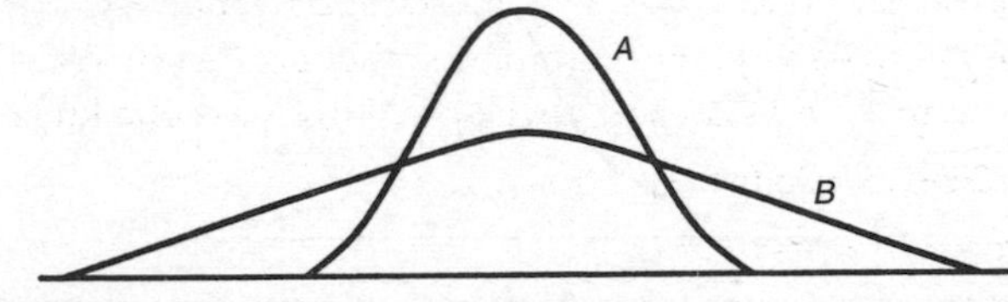

Fig. 36. **Dispersion.** Diagram of two distributions with the same mean, but different dispersions.

There are several measures of dispersion. Range is the difference between the largest and smallest members. By contrast, mean absolute deviation, standard deviation and semi-interquartile range are all measures of how far above or below the middle a typical member might be found. Standard deviation is the most important statistically. It gives more weight to extreme values than do the other two.

distribution the manner in which the values of a variable occur. This may be described in terms of their frequency or their probability. The variable may be discrete or continuous.

Example: *frequency distribution*. The lengths of reigns of English kings and queens, 827-1952 AD.

Length (y, years)	$y<10$	$10 \leq y<20$	$20 \leq y<30$	$30 \leq y<40$
Frequency	22	16	11	7

Length (y, years)	$40 \leq y<50$	$50 \leq y<60$	$60 \leq y<70$
Frequency	1	3	1

Example: *probability distribution*. The distribution of different totals when two dice are thrown. The top line (below) represents the possible combined scores of two dice. The bottom line shows the probability of each combined score occurring.

Total score	2	3	4	5	6	7	8	9	10	11	12
Probability	$\frac{1}{36}$	$\frac{2}{36}$	$\frac{3}{36}$	$\frac{4}{36}$	$\frac{5}{36}$	$\frac{6}{36}$	$\frac{5}{36}$	$\frac{4}{36}$	$\frac{3}{36}$	$\frac{2}{36}$	$\frac{1}{36}$

Figs. 37, 38, 39, 40, 41(a) and (b) show the graphs of some common distributions.

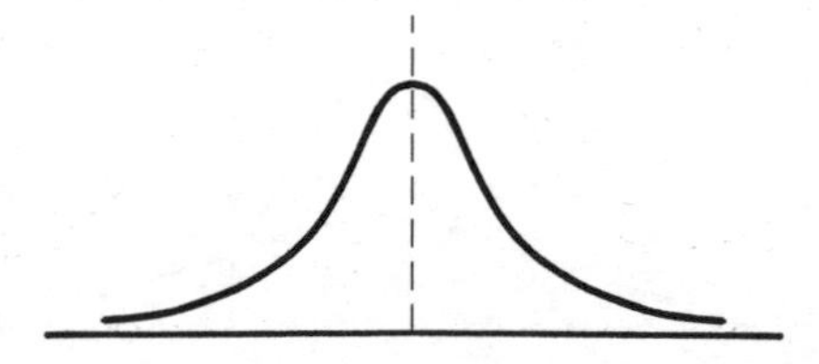

Fig. 37. **Distribution.** Normal distribution; the vertical line represents the mean value.

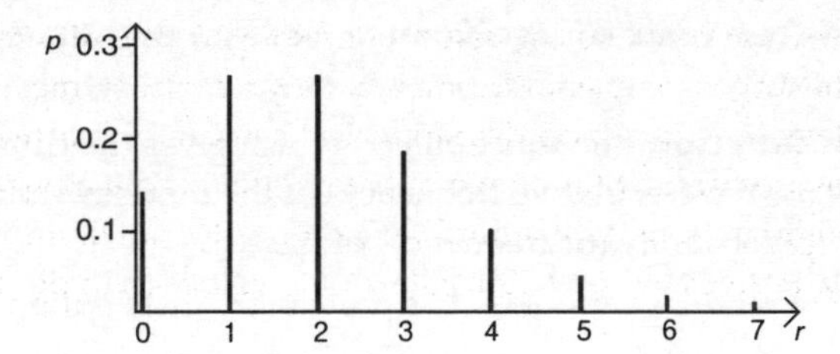

Fig. 38. **Distribution.** Poisson distribution; in this example, the mean μ is 2, so the distribution is denoted Poisson(2).

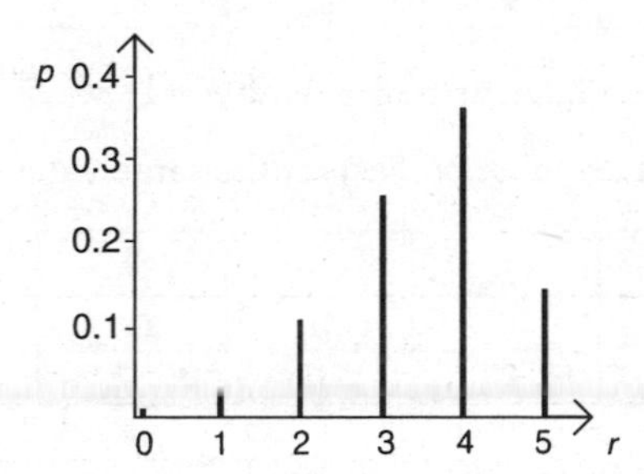

Fig. 39. **Distribution.** Binomial distribution; in this example, the number of trials n is 5 and the probability of success p is 0.7. Thus the distribution is denoted B(5, 0.7).

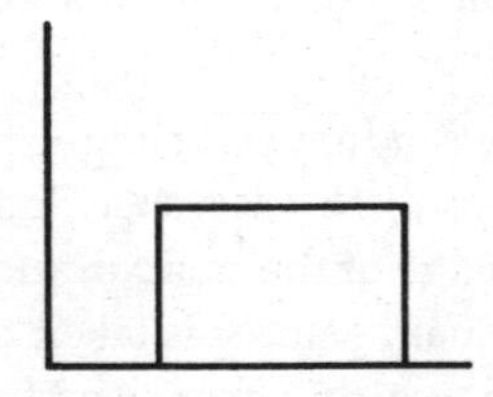

Fig. 40. **Distribution.** Continuous uniform (rectangular) distribution.

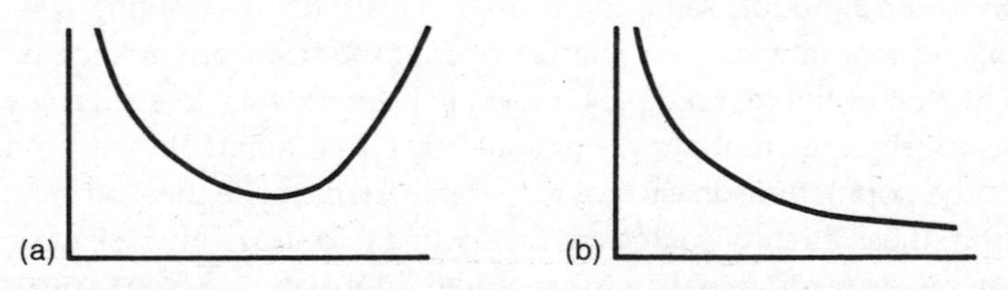

Fig. 41. **Distribution.** Empirical distributions described by their shapes. (a) U-shaped distribution. (b) J-shaped distribution.

distribution-free tests of significance see NON-PARAMETRIC TESTS OF SIGNIFICANCE.

distribution function the function F(x) which gives the cumulative probability (or the cumulative frequency) of the random variable X. Thus, F(x) = Probability (or frequency) of $X \leq x$.

Example: the random variable X has probability distribution p(x):

x	0	1	2	3	4	5
p(x)	$\frac{1}{32}$	$\frac{5}{32}$	$\frac{10}{32}$	$\frac{10}{32}$	$\frac{5}{32}$	$\frac{1}{32}$

(This is the binomial distribution with $p = \frac{1}{2}$, $n = 5$).

So the distribution function F(x) for the same values, is:

x	0	1	2	3	4	5
F(x)	$\frac{1}{32}$	$\frac{6}{32}$	$\frac{16}{32}$	$\frac{26}{32}$	$\frac{31}{32}$	$\frac{32}{32} = 1$

See CUMULATIVE FREQUENCY, also TABLE 2(a).

double blind experiment see BLIND EXPERIMENT.

double sampling a sampling procedure in which a preliminary sample is taken in order to obtain information that is needed in the design of the main survey.

Example 1: in a simple random sampling experiment, the sample size is to be chosen so as to obtain a specified precision in the results. This requires a knowledge of the standard deviation which is unknown. So a preliminary sample is taken to obtain an estimate of the standard deviation, which is then used to determine the total number of observations required, and so how many more are needed.

Example 2: double sampling is used in a stratified sampling experiment in which the relative sizes of the strata are unknown. The first of the two samples is used to estimate this. In such a case a relatively large first sample may be taken to estimate the proportions of the populations in each of the strata. Then smaller second samples from those already contacted, proportional to the relative sizes of the strata, are taken to obtain the required information. See STRATIFIED SAMPLING.

double sampling scheme (in quality control) a sampling scheme which may involve taking two samples. In such a scheme, a first sample is taken. If the number of defective items, D_1, is not more than the acceptance number, A_1, the whole batch is accepted. If it is greater than another, larger number, A_2, the batch is rejected (or subjected to 100% inspection). If the number of defectives lies between these two numbers, a second sample is taken and found to have D_2 defectives. If the total number of defectives between the two samples, $D_1 + D_2$, is then greater than a new acceptance number, A_3, the whole batch is rejected or subjected to a 100% inspection. Sometimes the new acceptance number, A_3, is the same as the larger number for the first stage of sampling, A_2, so $A_1 < A_2 \leqslant A_3$.

First sample: Number of defectives is D_1.

If $D_1 \leq A_1$, the batch is accepted.

If $A_1 < D_1 \leq A_2$, a second sample is taken.

If $D_1 > A_2$, the batch is rejected.

Second sample: Number of defectives is D_2.

If $D_1 + D_2 \leq A_3$, the batch is accepted.

If $D_1 + D_2 > A_3$, the batch is rejected.

See LATTICE DIAGRAM.

e

econometrics the application of mathematical and statistical techniques to economic theories.

effect variable see EXPLANATORY VARIABLE.

efficient estimator among two or more estimators, that one with the smaller variance. The use of this term is usually restricted to unbiased estimators.

Example: in an experiment, three independent readings X_1, X_2 and X_3 are taken of a random variable with mean μ and variance σ^2. Which of the following is the most efficient estimator for μ?

(a) $\dfrac{X_1 + X_2}{2}$ (b) $\dfrac{2X_1 + X_2 + X_3}{4}$ (c) $\dfrac{X_1 + X_2 + X_3}{3}$

All three are unbiased estimators since

(a) $$E\left(\frac{X_1 + X_2}{2}\right) = \frac{1}{2}E(X_1) + \frac{1}{2}E(X_2) = \frac{1}{2}\mu + \frac{1}{2}\mu = \mu$$

(b) $$E\left(\frac{2X_1 + X_2 + X_3}{4}\right) = \frac{2}{4}E(X_1) + \frac{1}{4}E(X_2) + \frac{1}{4}E(X_3)$$

$$= \frac{1}{2}\mu + \frac{1}{4}\mu + \frac{1}{4}\mu = \mu$$

(c) $$E\left(\frac{X_1 + X_2 + X_3}{3}\right) = \frac{1}{3}E(X_1) + \frac{1}{3}E(X_2) + \frac{1}{3}E(X_3)$$

$$= \frac{1}{3}\mu + \frac{1}{3}\mu + \frac{1}{3}\mu = \mu$$

Their variances are calculated as follows:

(a) $$\mathrm{Var}\left(\frac{X_1+X_2}{2}\right)=\mathrm{Var}\left(\frac{X_1}{2}\right)+\mathrm{Var}\left(\frac{X_2}{2}\right)$$

$$=\frac{1}{4}\mathrm{Var}(X_1)+\frac{1}{4}\mathrm{Var}(X_2)=\frac{1}{2}\sigma^2$$

(b) $$\mathrm{Var}\left(\frac{2X_1+X_2+X_3}{4}\right)=\mathrm{Var}\left(\frac{X_1}{2}\right)+\mathrm{Var}\left(\frac{X_2}{4}\right)+\mathrm{Var}\left(\frac{X_3}{4}\right)$$

$$=\frac{1}{4}\mathrm{Var}(X_1)+\frac{1}{16}\mathrm{Var}(X_2)+\frac{1}{16}\mathrm{Var}(X_3)$$

$$=\frac{3}{8}\sigma^2$$

(c) $$\mathrm{Var}\left(\frac{X_1+X_2+X_3}{3}\right)=\mathrm{Var}\left(\frac{X_1}{3}\right)+\mathrm{Var}\left(\frac{X_2}{3}\right)+\mathrm{Var}\left(\frac{X_3}{3}\right)$$

$$=\frac{1}{9}\mathrm{Var}(X_1)+\frac{1}{9}\mathrm{Var}(X_2)+\frac{1}{9}\mathrm{Var}(X_3)$$

$$=\frac{1}{3}\sigma^2$$

Since $\frac{1}{3}\sigma^2$ is the smallest variance, (c) is the most efficient of the three estimators.

eigenvalues, eigenvectors see TRANSITION MATRIX.

80-20 rule the observation that in many situations 80% of effects are the result of 20% of causes. The 80-20 rule is purely an empirical rule. It is a rough description of what is often observed to occur but it is not derived from any theory. The rule is applied in several areas of statistics. For example, when working with Pareto charts it gives a cut-off point for the number of possible sources of defects on which to focus attention. See PARETO CHART.

empirical derived from or relating to experiment and observation, rather than theory.

An example of an empirical law is the Rank Size Rule, used in geography. This states that if P_1 is the population of a country's largest city, then $\frac{P_1}{n}$ is that of its nth city. There is no theoretical basis for this rule; it has just been found to be approximately true in many cases as in the case of Japan (1995 figures).

Rank	*City*	*Actual size (millions)*	*Size predicted by the Rank Size Rule (millions)*
1	Tokyo	7.97	7.97
2	Yokohama	3.31	3.98
3	Osaka	2.60	2.66
4	Nagoya	2.15	1.99
5	Sapporo	1.76	1.59
6	Kyoto	1.46	1.33

An example of a law which is not empirical is that which states that when two fair dice are thrown repeatedly the possible scores 2, 3, 4, 5, 6, 7, 8, 9, 10, 11, 12, will occur with frequencies in the ratio.

1: 2: 3: 4: 5: 6: 5: 4: 3: 2: 1

This is based upon the fact that the probabilities of the different scores can be calculated. The probability of each entry in the table below is $\frac{1}{36}$; the number 8, for example, occurs 5 times and so the probability of scoring 8 is $\frac{5}{36}$.

First die						
6	7	8	9	10	11	12
5	6	7	8	9	10	11
4	5	6	7	8	9	10
3	4	5	6	7	8	9
2	3	4	5	6	7	8
1	2	3	4	5	6	7
	1	2	3	4	5	6
	Second die					

enumeration data data consisting of numbers of individuals in various categories.

Example: a flower has three varieties, red, white and pink.
400 flowers are collected and found to be:

Red	89
Pink	201
White	110

There are thus three categories, red, pink and white; the data consist of the number of flowers in each.

equiprobable of equal probability. The possible scores when a fair die is thrown, 1, 2, 3, 4, 5, and 6, are equiprobable, each has a probability of occurring of 1/6. The scores 0 to 36 on a fair French roulette wheel, are also equiprobable at 1/37 each.

By contrast the possible outcomes from tossing two coins (2 Heads; 1 Head 1 Tail; 2 Tails) are not equibrobable. Their probabilities are $\frac{1}{4}$, $\frac{1}{2}$ and $\frac{1}{4}$ respectively.

equivalent see SUPERIOR.

error **1.** a mistake or inaccuracy

2. the magnitude of an inaccuracy. See also RANDOM ERROR.

error curve a name sometimes given for the Normal distribution curve.

estimate **1.** in formal statistical use, the value of an estimator obtained from a particular sample.

2. in general use, an approximate calculation.

3. in general use, an approximate idea of something (e.g. size, cost).

4. to make an approximate calculation, or to form an approximate idea of something (e.g. size, cost).

The process of estimation is always based on information and/or reasoning. In this it differs from guessing which may have no rational basis.

estimator a random variable, formula or procedure for estimating the value of a parent population parameter from sample data. Its value depends on the particular sample involved, and is called an estimate. An estimator should ideally be consistent, unbiased and efficient. Estimators of population parameters are distinguished from their true values by the convention of using the symbol ^. See also EFFICENT ESTIMATOR, LINEAR ESTIMATOR.

Euler diagram see SET.

event something which may or may not occur. Examples: a coin showing heads when tossed; a particular seed germinating.

When a statistical experiment, like throwing a die, is carried out, the result is one of a number of possible outcomes, e.g. 1, 2, 3, 4, 5, 6. The term event is used to describe a combination of outcomes (e.g. the die shows an even number) or a particular outcome (e.g. the die shows 6).

exclusive (of two events) such that both events cannot occur. Thus events *A* and *B* are exclusive if *B* cannot occur when *A* does, and vice versa. It is, however possible that neither *A* nor *B* occurs.

Example: *A*: getting a 6 when throwing a die
B: getting a 4 on the same throw on the die

When three or more events are such that no two of them can occur at the same time, they are described as *mutually exclusive*. However this term is also often used for two exclusive events. Compare disjoint in SET.

exhaustive **1.** (of two events) such that at least one must occur. Thus events *P* and *Q* are exhaustive if either *P* alone, or *Q* alone, or both *P* and *Q* together, *must* occur; it is not possible for neither to occur. Exhaustive events cover all possibilities.

Example: Jayne is an athlete
P: Jayne does a track event
Q: Jayne does a field event

Since all athletics events are track, or field, or both (e.g. the decathlon), *P* and *Q* are exhaustive events.

2. (of a greater number of events) such that they cover all possibilities between them; they are comprehensive in their scope.

expectation the expected or mean value of a random variable, or of a function of the variable. Symbol: E( ). For a discrete variable, *X*, taking values $x_1, x_2, \ldots, x_n$ (where n may be infinite) with probabilities $p(x_1), p(x_2), \ldots, p(x_n)$, the expectation of *X*, E(*X*) or μ is given by:

$$E(X) = \sum_{i=1}^{i=n} x_i p(x_i)$$

and for a continuous variable

$$E(X) = \int x f(x) dx$$

where f(x) is the probability density function of *X*.

Example: when a die is thrown, each of the possible values 1, 2, 3, 4, 5 and 6 has probability of 1/6 of showing. So the expectation of the value shown is:

$$\left(1\times\frac{1}{6}\right)+\left(2\times\frac{1}{6}\right)+\left(3\times\frac{1}{6}\right)+\left(4\times\frac{1}{6}\right)+\left(5\times\frac{1}{6}\right)+\left(6\times\frac{1}{6}\right)=3.5$$

Expectation obeys the following laws:

$$E(kX) = k(X) \text{ where } k \text{ is a constant}$$

$$E(X_1 \pm X_2) = E(X_1) \pm E(X_2)$$

$$E(X^2) = [E(X)]^2 + \text{Var}(X)$$

$$E(X_1X_2) = E(X_1)E(X_2) + \text{Cov}(X_1, X_2)$$

where Var is variance and Cov is covariance.

If g(X) is a function of the random variable X with probability density function f(x), then

$$E\left(g(X)\right) = \int g(x)f(x)dx$$

so that, for example,

$$E(X^m) = \int x^m f(x)dx$$

$$E(\sin X) = \int \sin x \, f(x)dx$$

experiment any process which results in the collection of data. This meaning for the word is somewhat different from that usually given to it in other scientific disciplines.

Examples of experiments in a statistical sense include: carrying out a survey; measuring the daily rainfall; recording the longevity of animals of a particular type.

experimental design the way in which a statistical EXPERIMENT is planned. The example which follows illustrates the use of a number of different designs, namely: *completely randomised design, randomised blocks design, Latin square* and *Graeco-Latin square.*

Example: a market gardener wants to test three types of peas, *A, B* and *C*, on his land. He has a square plot which he divides into nine equal squares, three to be planted with each type of pea. The problem which he then faces is which square to plant with which type.

One method is complete randomisation which might, for example, result in the pattern in Fig. 42

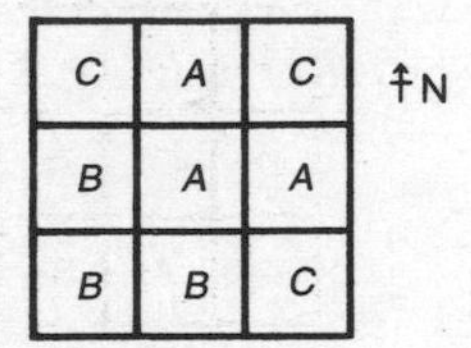

Fig. 42. **Experimental design.** Completely randomized design for a square with nine equally-sized plots, in a test for three types of pea, *A, B, C.*

This would be all right if all the plots were equally desirable. If however, there were a prevailing north wind so that the northernmost plots were more exposed, he might decide to use randomised blocks (Fig. 43), where each of the types *A, B* and *C* is planted once in each west-east block.

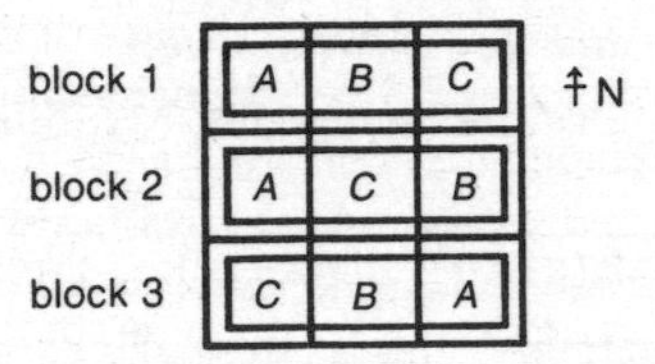

Fig. 43. **Experimental design.** A square plot divided for a randomized block design for three types of pea, *A, B, C.*

If the gardener also felt that the soil to the east was rather better than that to the west, he would use a Latin square (Fig. 44), where each type of pea is planted once in each row (west-east), and once in each column (north-south).

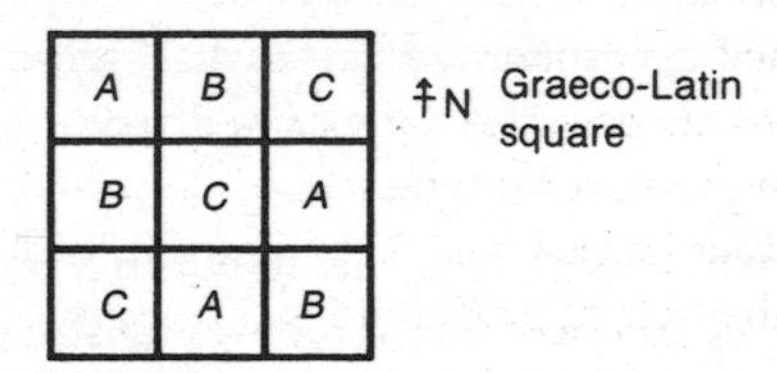

Fig. 44. **Experimental design.** A square plot divided into a Latin square, to test three types of pea, *A, B, C,* where each type appears once in each row, and once in each column.

If he also wished to eliminate the effects of a third variable, for example the use of three insecticides, α, β, and γ, he would use a Graeco-Latin square (Fig. 45). In this experimental design, each type of pea appears once in each row, once in each column and once with each of the three types of insecticide.

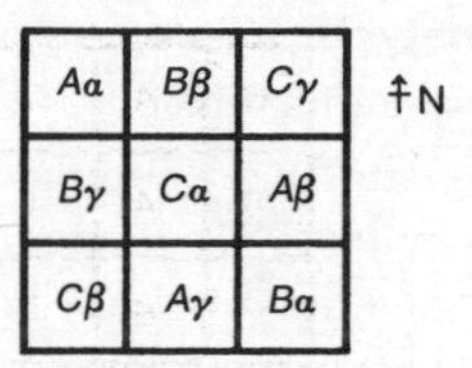

Fig. 45. **Experimental design.** A square plot divided into a Graeco-Latin square to eliminate the effects of a third variable, α, β, γ.

A Graeco-Latin square is really two Latin squares, one superimposed on the other. There are Graeco-Latin squares of sides 3, 4, 5, 7, 8, 9 and 10, but not 6.

experimental error the difference between the value of a quantity found in an experiment and that which should have been found had the experiment been conducted perfectly. Experimental error is due to experimental technique, and does not include mistakes in mathematics or experimental design. No matter how carefully an experiment is conducted, experimental error is always present, for example because of small changes to the ambient conditions, and it is the role of statistics to measure, analyse and allow for the error.

Example: a physics student conducting an experiment to find the gravitational acceleration *g*, gets an answer of 9.99 ms^{-2} instead of 9.81 ms^{-2}.

The *experimental error* is 0.18ms^{-2}.

The *percentage experimental error* is $\frac{0.18}{9.81} \times 100\% = 1.83\%$.

experimental or **empirical probability** a probability which is determined by experiment or observation rather than from some underlying theory. The figures used by life insurance companies in calculating risks for particular individuals come into this category. On the other hand, a skilled bridge player decides strategy on the basis of theoretical probabilities.

explanatory variable 1. or **cause variable** a variable upon which another depends. If one variable *Y* is caused by another *X*, then *X* is called the explanatory variable, and *Y* the *effect variable*. If, for example, cigarette smoking is regarded as causing lung cancer, then the number of cigarettes smoked per day could be the cause variable, the probability of death by lung cancer the effect variable.

2. or **independent variable** (see DEPENDENT VARIABLE). In this case the idea of cause may or may not be involved.

exploratory data analysis the process of looking at raw data to decide on their important features.

The term is linked to John Tukey and the collection of techniques which he developed in the 1970s. It refers to the initial analysis of data to see what information can easily be found from them, and is seen as a counter balance to too much emphasis on hypothesis tests being done without regard to the data themselves.

Exploratory data analysis may involve such procedures as:

(a) Rounding the figures, or cutting out unnecessary numbers of significant figures

(b) Grouping the data in a convenient form, such as a stem-and-leaf plot, a histogram or a box-and-whisker plot. Box-and-whisker plots are particularly useful in comparing the central tendency and spread from more than one batch

(c) Identifying outliers

(d) Finding the median and quartile values

(e) More sophisticated techniques like median polish and the use of a resistant line.

The techniques often use statistics which are "resistant" to the effects of outliers. So, for example, they use the median rather than the mean, as in the case of median polish rather than sweeping by means for the analysis of a two-way table. Similarly the resistant line is an alternative to the least squares regression line, which is resistant to outliers.

exploratory survey see PILOT SURVEY.

exponential (of a value or function) raised to the power of.

Thus the function $f(x) = a^x$ is an exponential function: x is called the *exponent*.

The term exponential is most commonly used when the exponent is the power of e, the base of natural logarithms. In this case the exponential of x is e^x, which is sometimes written as $\exp(x)$.

exponential distribution a continuous distribution (see Fig. 46) with probability density function given by:

$$f(x) = \lambda e^{-\lambda x}$$

For this distribution,

$$Mean = \frac{1}{\lambda} \qquad Variance = \frac{1}{\lambda^2}.$$

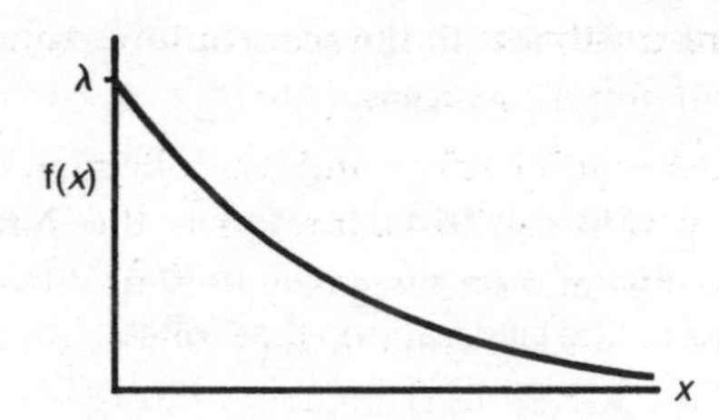

Fig. 46. **Exponential distribution.**

The exponential distribution is, among other things, sometimes used to model the life expectancy of materials.

The moment generating function of this distribution is $M(\theta) = \frac{\lambda}{\lambda - \theta}$.

See MOMENT GENERATING FUNCTION.

exponential smoothing a technique employed when using time series data to forecast future values.

Example: the value of the monthly sales of a product over the last year are given in this table.

Month	Jan	Feb	Mar	Apr	May	June	July	Aug	Sep	Oct	Nov	Dec
Sales (×£1000)	2.4	2.5	2.1	2.7	3.1	3.1	3.0	3.5	3.2	3.7	3.4	3.3

The company wishes to forecast the sales in the coming January.

If the values in a time series are independent, their mean, in this case 3.0, is the most useful piece of information for forecasting the next value. However, in this case the values do not seem to be independent; there appear to be runs of low values, of middle values and of high values. It seems likely that recent values will give better information about what the next value is likely to be.

Single exponential smoothing is a method for forecasting values where the values in a time series are not independent, but there is neither trend nor seasonal pattern.

If the values recorded so far are $y_1, y_2, \ldots y_n$ then the estimate of the next value is given by a weighted mean of the most recent value (weight α) and the estimate (weight $1-\alpha$) which had previously been obtained of that value:

$$\hat{y}_{n+1} = \alpha y_n + (1-\alpha)\hat{y}_n \text{ where } 0 \leqslant \alpha \leqslant 1$$

In other words, the estimate of the next value is somewhere between where the current value was forecast to be and where it turned out to be.

To start the calculation, it is usual to assume that the initial value y_1 and its forecast value $\hat{y}_1$ were the same, in this case 2.4. A value of α is also required and in the calculations that follow it is taken (arbitrarily) that $\alpha = 0.3$.

The forecast value for y_2 is then given by

$\hat{y}_2 = \alpha y_1 + (1-\alpha)\hat{y}_1 = 0.3\times2.4+0.7\times2.4 = 2.4$

Similarly the forecast value for y_3 is found as

$\hat{y}_3 = \alpha y_2 + (1-\alpha)\hat{y}_2 = 0.3\times2.5+0.7\times2.4 = 2.43$,

and so on.

The data can now be presented as a table of previous values and the forecasts that were made for them at the relevant time.

Month, i	1	2	3	4	5	6	7	8	9	10	11	12
Sales, y_i	2.4	2.5	2.1	2.7	3.1	3.1	3.0	3.5	3.2	3.7	3.4	3.3
Forecast, $\hat{y}_i$	(2.4)	2.4	2.43	2.33..	2.44..	2.63..	2.77..	2.84..	3.04..	3.08..	3.27..	3.31..

Most of the numbers in the Forecast row have been truncated and this is indicated by the use of "..". Thus the final figure, $\hat{y}_{12}$, is actually 3.31044 to 5 decimal places, or £3310.44 to the nearest 1p.

The forecast for next January is now calculated as

$\hat{y}_{13} = \alpha y_{12} + (1-\alpha)\hat{y}_{12} = 0.3\times3.3+0.7\times3.31044.. = 3.30731..$

If the relationship between successive values is very strong, it is appropriate to place a lot of emphasis on the most recent value and so to use a large value of α. If the relationship is less strong, more emphasis is placed on the less recent values and so a smaller value of α is used.

If small values of α are being used, a lot of emphasis is given to the initial value, and in this situation a new starting value is introduced, y_0^*, which is simply the mean of the values in the series. Otherwise, it is usual to use $\hat{y}_2 = y_1$ as above.

The use of a computer makes it possible to investigate the effects that different values of α would have had on the accuracy of forecasting the values in the time series up to the current point, and so to choose the "best" value (according to some suitable criterion) for the particular situation.

Similar methods (double and treble exponential smoothing) exist for use in the cases where there is a trend and / or seasonal factors.

extrapolation estimation of a value of a variable beyond known values. Extrapolation is a dangerous process which may lead to false results, as illustrated in the following example.

Example: A girl was 1.20 m tall on 1 January 2003, and 1.40 m on 1 January 2008. Estimate her height on 1 January 2013.

The graph in Fig. 38 uses extrapolation. It gives the impression that by 1 January 2013, she would have grown another 0.20 m to be 1.60 m tall. This assumes that she continued to grow at the same rate and may or may not be true. Eventually, however, this assumption must become false; otherwise she would keep on growing and growing and, for example, by 1 January 2036 she would be a record-breaking giantess of 2.52 m.

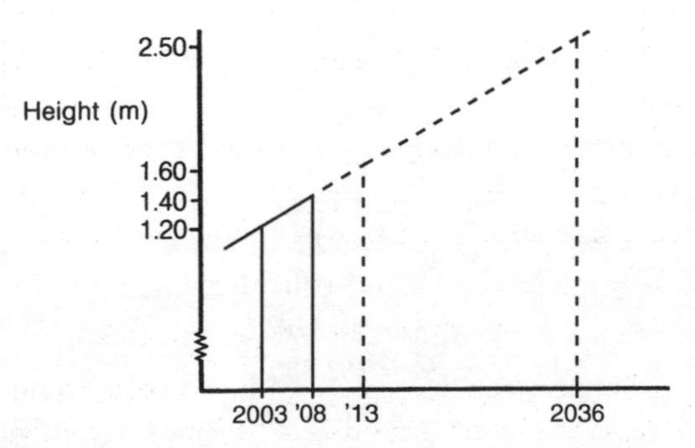

Fig. 47. **Extrapolation.** Graph of a girl's height in 2003 and 2008, with her height extrapolated for 2013 and 2036.

Extrapolation may be reasonably accurate near the known points of information, but the further away from them one works, the less reliable the estimate becomes.

Compare INTERPOLATION.

factor analysis a collection of techniques used in multivariate analysis mainly to reduce the number of variables and to find structure in the relationships between the variables.

factorial the product of all the positive integers from 1 up to and including a given integer. Symbol: $n!$ where n is the given integer. Thus 5 factorial, written 5!, means

$$1\times2\times3\times4\times5 \qquad (=120)$$

By convention, 0! is given the value 1.

Factorials are used in working out probabilities in the Poisson distribution and binomial distribution, in permutations and combinations, etc.

Factorials very quickly become large numbers.

$$1! = 1$$
$$2! = 2$$
$$3! = 6$$
$$4! = 24$$
$$5! = 120$$
$$6! = 720$$
$$7! = 5\,040$$
$$8! = 40\,320$$
$$9! = 362\,880$$
$$10! = 3\,628\,800$$

For large values of n,

$$n! \approx \sqrt{2\pi n}\, n^n e^{-n}, \text{ or } n! \approx \sqrt{2\pi}\, n^{(n+\frac{1}{2})} e^{-n}$$

which can be rewritten as *Stirling's approximation* to $n!$

$$\log_{10} n! = 0.39909 + \left(n + \frac{1}{2}\right) \log_{10} n - 0.4342945n$$

factorial experiment an experiment designed to provide information about the effects of individual factors which may be used in combination, and the interactions between them. A factorial design is very powerful and yields a lot of information.

Example: when sealing a bag of crisps, the pressure applied, the temperature and the duration are all relevant to the strength of the seal which is obtained. It is known from a pilot experiment that it is appropriate just to investigate a "low" and a "high" value for each of these. Since there are three factors, each with two levels, there are 2^3 possible combinations and the investigation is called a 2^3 factorial experiment. (If there were two factors, one with 3 levels and the other with 5 levels, it would be a 3×5 factorial experiment, with 15 possible combinations.)

Suppose these factors are denoted by the letters A, B and C.

A combination of a level, "high" or "low", for each factor is known as a treatment. Various notations are used to indicate the levels in a treatment combination.

The most useful notation is to let the lower case letter, a, indicate that factor A is at its high level; the absence of letter a indicates that A is at its low level. Similarly for the other factors. Thus ac indicates that A and C are at their high levels and that B is at its low level. If all the factors are at their low levels, the notation (1) is used; it is read simply as "one". (This notation cannot of course be used if factors have more than 2 levels; in such cases notations such as a_1, a_2, a_3 are used.) Thus the 8 possible treatment combinations for A, B and C are:

$$(1), a, b, ab, c, ac, bc, abc.$$

Consider, for example, factor C. Notice that the first 4 treatment combinations in this list have c at a low level (the letter c is absent) whereas the last 4 have C at its high level (the letter c is present). So the overall effect of C, called the *main effect of* C, can be measured by subtracting the measurements for the first 4 treatment combinations from those from the last 4. (It is common practice to divide this by 4 so that the measure is given as a mean difference.) This can be done similarly for the other factors.

The principle is the same in any experiment where all the factors have only two levels. If there are factors with more than two levels, the principle is extended appropriately.

2^2 and 2^3 factorial experiments may be visualised as shown in Fig. 48(a) and (b).

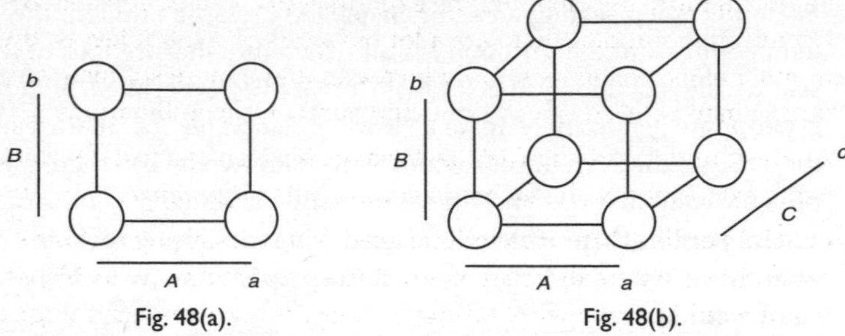

Fig. 48(a). Fig. 48(b).

Fig. 48. **Factorial experiment**, shown diagramatically for 2 factors (a) and for 3 factors (b).

The yields from each of the treatment combinations may be put in the circles and the effects, i.e. the differences, written on the lines joining the circles. The main effect of factor *A* is then found as the average of the values on the horizontal bars, and similarly for the other factors.

Another important feature is the possibility of interaction between the factors.

Consider first a simple 2^2 experiment with 2 factors each at 2 levels, for example pressure and temperature affecting the yield of the product of a chemical reaction. Perhaps it is best, on the whole, to use the "high" level of the pressure; so pressure has a positive main effect. Perhaps it is also best, on the whole, to use the "high" level of temperature; so temperature has a positive main effect. However, it may be that the particular combination of "high" pressure and high temperature causes the yield of the product to decrease; this might, for example, happen because at these "high" pressures and temperatures some other by-product of the chemical reaction takes over. Alternatively, or as well, the particular combination of "low" pressure might cause the yield of the product to decrease, perhaps because the reaction as a whole only goes rather slowly when both pressure and temperature are "low". In cases like these examples, the factors are said to *interact*.

Interaction between factors can occur in a factorial experiment of any size. If there are several factors, there could be two-factor interactions, three-factor interactions, four-factor interactions, and so on.

If the interactions are large it can be misleading to quote the main effects in isolation.

To obtain more information, the experiments could be replicated. On the other hand, if there are a lot of factors or a lot of levels, the number of experiments rapidly increases and a more sophisticated design may be used which limits the number of combinations.

Analysis of data arising from factorial experiments is usually done using extensions of the technique of analysis of variance.

factorial replication a technique used when designing factorial experiments where there are many factors and / or many levels, so that it would be unwieldy or very expensive (or even technologically impossible) to investigate all possible treatment combinations. The aim is to investigate only a fraction of the possible combinations (often a simple fraction such as $\frac{1}{2}$, $\frac{1}{4}$ or $\frac{1}{8}$) chosen in such a way that good information about the effects of the factors can still be obtained in the analysis.

Example: A 2^3 experiment factorial experiment has three factors, A, B and C. There are 8 possible treatment combinations but only enough resources to investigate 4 of them, i.e. one half of the total. It is not, however, sensible to choose the 4 treatment combinations in which C is at its "low" value; it would then be impossible to obtain any information at all about the overall (main) effect of factor C.

***F* distribution** the distribution underlying the F test. Critical values are given in Table 5. The F distribution is the distribution of the random variable that is the ratio of two independent chi-squared random variables each divided by its degrees of freedom. Notation $F_{m,n}$. Thus

$$F_{m,n} = \frac{\dfrac{\chi^2_m}{m}}{\dfrac{\chi^2_n}{n}}$$

where the two chi-squared random variables are independent.

From this it follows that it is also the distribution of the ratio of the sample variances of independent random samples drawn from two independent Normal populations with the same population variance.

The distribution has two degrees of freedom parameters. It is extensively used in statistics, for example in the F test and in analysis of variance. See F TEST.

fertility rate the ratio of live births in a specified area, group, etc., to the female population between the ages of 15 and 44, usually expressed per thousand per year.

finite population correction a correction factor applied to the expression for the variance of the means of samples drawn from an infinite parent population, to allow for the population being of finite size. When samples of size n are taken from an infinite parent population with variance σ^2, the variance of the sample means is given by:

$$Variance = \frac{\sigma^2}{n}.$$

If, however, the samples are taken from a finite population of size N, without replacement, the variance and standard deviation of the sample means are given by:

$$Variance = \frac{(N-n)}{N(N-1)}\sum_{i=1}^{N}(x_i-\bar{x})^2$$

The population variance for a finite population is often calculated as

$$\sigma^2 = \frac{1}{N-1}\sum_{i=1}^{n}(x_i-\bar{x})^2.$$

In this case, the variance of the sample means is given by

$$Variance = \frac{(N-n)}{N}\sigma^2 = \left(1-\frac{n}{N}\right)\sigma^2.$$

This expression for the variance differs from that for an infinite population by the factor $\left(1-\frac{n}{N}\right)$ and this is called the finite population correction factor. It is also written $(1-f)$ where f is the sampling fraction.

The population variance for a finite population is, alternatively, sometimes calculated as

$$\sigma^2 = \frac{1}{N}\sum_{i=1}^{n}(x_i-\bar{x})^2.$$

In this case, the variance of the sample means is given by

$$Variance = \frac{(N-n)}{N-1}\sigma^2$$

and the finite population correction factor is given by $\frac{N-n}{N-1}$.

If N is large in comparison with n, the finite population correction may be ignored.

Fisher's Ideal Index see INDEX NUMBER.

Fisher's *z* transformation a transformation applied to the product moment correlation coefficient in testing the null hypothesis that the population correlation coefficient has a particular value. It is also used in setting up confidence limits for the value of the population correlation coefficient.

Fisher's z transformation transforms the distribution of the sample correlation coefficient, r, into a Normal distribution. The new variable, z, is given by

$$z = \frac{1}{2}\ln\left(\frac{1+r}{1-r}\right) = \text{artanh}\, r.$$

Alternatively Table 9 may be used.

For samples of size n, drawn from a bivariate population with correlation coefficient ρ, the distribution of z is Normal with mean

$$\frac{1}{2}\ln\left(\frac{1+\rho}{1-\rho}\right) \text{ and variance } \frac{1}{n-3}.$$

It is often the case that the null hypothesis for a test on the correlation coefficient is that there is no correlation between the two variables so that $\rho = 0$. Such a test is usually carried out as described under CORRELATION COEFFICIENT, and Table 6 makes it unnecessary to use Fisher's z transformation. There are however times when a test is required for the null hypothesis that ρ has some other value, and it is then that it is appropriate to use this technique.

Example: it is suggested that there is a linear association between the incidence of fleas and tapeworms in domestic cats, and that the correlation coefficient between the number of fleas on a cat and the number of tapeworms is 0.4. In an experiment 100 cats were examined; the correlation coefficient for these two random variables was found to be 0.25 for this sample. Carry out a hypothesis test, at the 5% significance level, on the suggested value for the population correlation coefficient.

H_0 $\rho = 0.4$: The sample is drawn from a population with correlation coefficient 0.4.

H_1 $\rho \neq 0.4$.

5% significance level

2-tail test

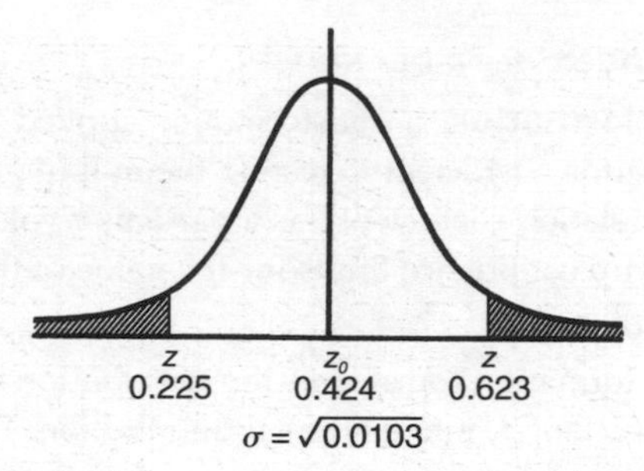

Fig. 49. **Fisher's z transformation.** Critical region for the hypothesis test in the example.

Using Fisher's transformation, the sampling distribution for the transformed correlation coefficients of samples of size 100, assuming H_0 to be true, is

$$N\left(\frac{1}{2}\ln\left(\frac{1+0.4}{1-0.4}\right), \frac{1}{100-3}\right) = N(0.424, 0.0103).$$

So for the test in question, the critical values are

$0.424 \pm 1.96 \times \sqrt{0.0103}$, or 0.225 and 0.623 (see Fig.48)

The test statistic is the value of z corresponding to the sample correlation coefficient of 0.25. This is given by:

$$z = \frac{1}{2}\ln\left(\frac{1+0.25}{1-0.25}\right) = 0.255$$

Since 0.255 > 0.225, the null hypothesis is accepted.

Fisher's z transformation is also used to find confidence intervals for the population correlation coefficient. The sample data in the example above could be used in this way, rather than for a hypothesis test. The 95% confidence limits for z would be $0.255 \pm 1.96 \times \sqrt{0.0103}$ = 0.056 or 0.454. These are transformed back into values for the correlation coefficient using the inverse transformation $r = \tanh z$, to obtain the confidence limits for ρ as 0.053 and 0.428.

Confidence limits for ρ may also be found using charts such as that in Fig. 50. In this example, r is 0.25, and the sample size n is 100. On the chart, the line, $r = 0.25$ intersects both curves for $n = 100$; the values of ρ corresponding to these two intersections are seen to be 0.056 and 0.454, the confidence limits for the population correlation coefficient. See CORRELATION COEFFICIENT, CONFIDENCE INTERVAL.

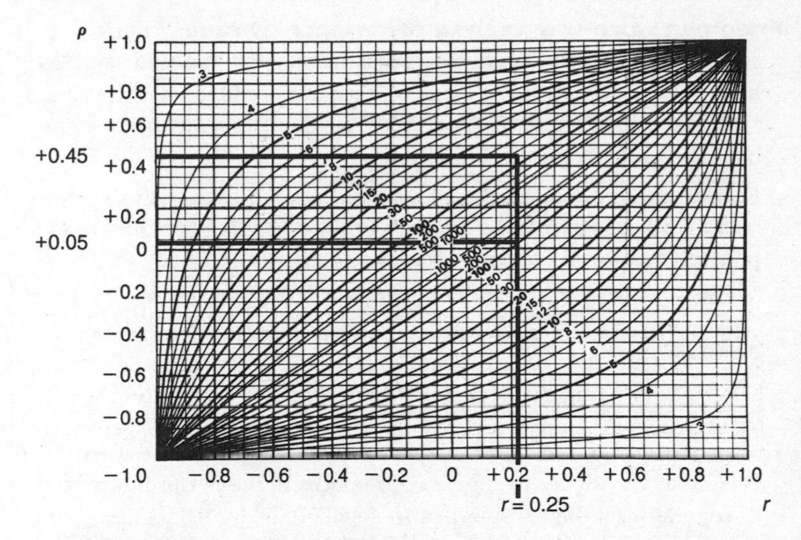

Fig. 50. **Fisher's z transformation.** 95% confidence interval chart for correlation coefficients. Source: *Elementary Statistics Tables*, H.R. Neave, George Allen and Unwin.

frame a representation of the items available to be chosen for a sample. This could be a telephone directory, a map, an electoral register, a list of the cattle in a herd, etc.

frequency **1.** or **absolute frequency** the number of individuals in a class. Symbol: f.

Example: in 20 shots at a target, a marksman makes the following scores:

4, 3, 2, 4, 5, 2, 4, 2, 5, 1, 0, 5, 1, 5, 2, 5, 5, 0, 1, 4

The frequencies of the different scores are thus:

Score	0	1	2	3	4	5
Frequency	2	3	4	1	4	6

2. See RELATIVE FREQUENCY.

frequency distribution see DISTRIBUTION.

frequency surface a three-dimensional representation of a SCATTER DIAGRAM. The area of the graph is divided into a grid, each part of which is raised to a height proportional to the frequency of the points lying within it.

Friedman's two-way analysis of variance by rank a non-parametric test of the null hypothesis that several matched samples have been drawn from the same underlying population. This is a 1-tail test with the alternative hypothesis that the samples have not been drawn from the same population.

Example: four operators (Op. 1, 2, 3 and 4) used each of three screw-making machines for a fixed length of time; the numbers of screws produced were:

	Op. 1	*Op. 2*	*Op. 3*	*Op. 4*
Machine A	361	385	340	377
Machine B	262	283	358	259
Machine C	420	434	392	445

Is there any evidence to suggest differences in the performances of the operators at the 5% significance level?

H_0 There is no difference in the operators' performance.

H_1 There is a difference in the operators' performance.

5% significance level

1-tail test

The figures in each row of the table are ranked (1 to 4) for the four operators, and the sums of the ranks for each calculated (R_i).

	Op. 1	*Op. 2*	*Op. 3*	*Op. 4*
Machine A	3	1	4	2
Machine B	3	2	1	4
Machine C	3	2	4	1
Total (R_i)	9	5	9	7

The value of the test statistic M is then worked out using the formula:

$$M = \frac{12}{NK(K+1)} \sum_{i=1}^{K} R_i^2 - 3N(K+1)$$

where N is the number of machines (3), K is the number of operators (4), and R_i is the total of the ranks for operator i. Thus

$$M = \frac{12}{3\times4\times(4+1)}(81+25+81+49) - 3\times3\times(4+1) = 2.2$$

The critical value for this statistic is found in the χ^2 tables (Table 4). There are $v = K - 1$ (= 3) degrees of freedom.

At the 5% level, the critical value for 3 degrees of freedom is 7.81.

Since 2.2 < 7.81, there is no reason to reject the null hypothesis.

There is no evidence of a significant difference among the operators' performance. See also CHI-SQUARED DISTRIBUTION.

***F* test** a test that two independent samples have been drawn from populations with the same variance. It is based on comparing the sample variances. The *F* test requires the underlying populations to be Normally distributed, but it is sometimes a reasonable approximation when this is not the case.

If two samples of sizes n_1 and n_2 are taken from populations with the same variance and are used to give estimates, s_1^2 and s_2^2, of the population variance, those two estimates should be approximately equal,

particularly if n_1 and n_2 are large. If the ratio $\frac{s_1^2}{s_2^2}$ is not close to 1, then there may be reason to suspect that the two samples are drawn from populations with different variances.

The test statistic *F* is given by

$$F_{v_1,v_2} = \frac{s_1^2}{s_2^2}$$

where $v_1 = n_1 - 1$ and $v_2 = n_2 - 1$ are the degrees of freedom.

F_{v_1,v_2} has the *F* distribution with v_1 and v_2 degrees of freedom.

The *F* distribution tables (Table 5) list critical values for F_{v_1,v_2} for different significance levels and degrees of freedom. The sample with the greater variance is designated number 1.

If the value of the test statistic is greater than the listed critical value, then the null hypothesis (that the samples are drawn from populations with the same variance) is rejected. This is a two-tail test (see ONE- AND TWO-TAIL TESTS), and so the percentage point to be looked up in the tables is half of the required significance level.

Example: two scientists, Dr MacDonald and Dr Desai, find a previously unknown type of fish in a remote river. They both trap some fish, from different locations some distance apart. The weights of their fish are as follows (in kg):

Dr MacDonald	0.15, 0.18, 0.25, 0.36, 0.42, 0.44	$n = 6$
Dr Desai	0.25, 0.26, 0.26, 0.30, 0.32, 0.33, 0.37, 0.37	$n = 8$

Do the figures support, at the 10% significance level, the theory that the fish came from populations with the same variance?

The test is

$H_0 \quad \sigma_1^2 = \sigma_2^2$

$H_1 \quad \sigma_1^2 \neq \sigma_2^2$

10% significance level

2-tail test

For Dr MacDonald's fish, $s^2 = 0.0154$

For Dr Desai's fish, $s^2 = 0.00234$

So Dr MacDonald's fish, having the greater variance, are called sample 1.

$$n_1 = 6 \qquad s_1^2 = 0.0154$$
$$n_2 = 8 \qquad s_2^2 = 0.00234$$

$$F = \frac{s_1^2}{s_2^2} = \frac{0.0154}{0.00234} = 6.58$$

The degrees of freedom are:

$$v_1 = 6 - 1 = 5 \qquad v_2 = 8 - 1 = 7$$

Since the *F* test is 2-tail, the percentage point to be looked up is $\frac{1}{2} \times 10\% = 5\%$.

From the *F* distribution tables (Table 5) for 5% points, the critical value for $F_{5,7}$ is 3.97.

Since 6.58 > 3.97, the null hypothesis is rejected. So the evidence suggests that, at the 10% significance level, the fish caught by the two scientists did not come from a common population.

Since the null hypothesis for the *F* test on the variances has been rejected, there is no point, or indeed validity, in applying a standard *t* test on the differences in the sample means. The *t* test assumes that both populations have a common variance.

function a relation between two sets that associates a unique element of the second with any element of the first.

Examples:

$P = 1000e^{0.001t}$; P is a function of t

$y = 3+\sin 2x$; y is a function of x.

The notation $y = f(x)$ reads, "y is a function of x".

A function must be unique for any value of its variable. Thus $y = \pm\sqrt{x}$ is not a function, since, for example, when $x = 9$, $y = +3$, or -3 and so the value of y is not unique.

g

gamma distribution the distribution with probability density function f(x) given by:

$$f(x) = \frac{\lambda^{\alpha}}{\Gamma(\alpha)} x^{\alpha-1} e^{-\lambda x} \text{ for } x \geq 0$$

where α and λ are parameters of the distribution, and $\Gamma(\alpha)$ is the gamma function, given by:

$$\Gamma(\alpha) = \int_0^{\infty} x^{\alpha-1} e^{-x} dx \text{ (for } \alpha \text{ real and positive).}$$

If α is a positive integer, $\Gamma(\alpha) = (\alpha - 1)!$

The gamma distribution has

$$Mean = \frac{\alpha}{\lambda}, \; Variance = \frac{\alpha}{\lambda^2}$$

The moment generating function of the gamma distribution is $\left(\frac{\lambda}{\lambda - t}\right)^{\alpha}$.

The graph of the gamma distribution varies in shape according to the value of α, as shown in Fig. 51

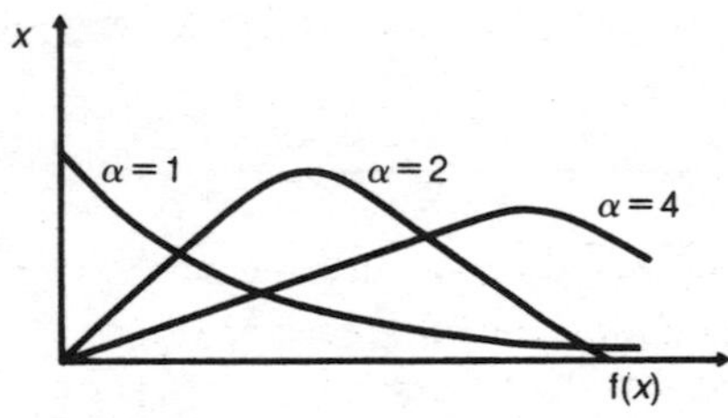

Fig. 51. **Gamma distribution.** The different curves for values 1, 2, 4 of the parameter α.

The gamma distribution provides a very flexible model for non-negative continuous random variables.

In the case $\alpha = 1$, the gamma distribution is the *exponential distribution*.

In the case $\alpha = \frac{k}{2}$ and $\lambda = \frac{1}{2}$, it gives the chi squared distribution, χ_k^2.

Gaussian curve see NORMAL CURVE.

geometric distribution the probability distribution $p(r)$ given by:

$$p(r) = pq^{r-1} \quad \text{for } r = 1, 2, \ldots \text{ where } 0 < p < 1, q = 1 - p$$

It represents the probability of the first success in a series of Bernoulli trials occurring at the rth attempt. Each trial has probability p of success, probability $q = 1 - p$ of failure.

For this distribution

$$Mean = \frac{1}{p}, \; Variance \; \frac{q}{p^2}$$

Example: a die is thrown until a six occurs. What is the probability that the number of throws is 5?

The distribution is geometric, with

$$p = \frac{1}{6} \qquad q = \frac{5}{6}$$

So the probability that the number of throws is 5 is:

$$\frac{1}{6} \times \left(\frac{5}{6}\right)^4 = 0.0804\,.$$

Compare PASCAL'S DISTRIBUTION.

geometric mean the geometric mean of a set of n numbers is the nth root of their product. Thus, the geometric mean of the n numbers $x_1, x_2, \ldots x_n$ is:

$$\sqrt[n]{x_1 x_2 \ldots x_n}$$

An example of the use of a geometric mean is in working out an "average" inflation rate. The inflation rates in the UK (retail prices index) for the years 1977-81 were:

1977	*1978*	*1979*	*1980*	*1981*
15.8%	8.3%	13.4%	18.0%	11.9%

The average inflation rate is that which, applied uniformly over the same period, would have the same overall effect. To work this out, it is first necessary to see the inflation figures as scale factors by which the cost of living in increased, namely:

1977	*1978*	*1979*	*1980*	*1981*
1.158	1.083	1.134	1.180	1.119

Thus, goods which cost £1 at the start of 1977 cost £1 × 1.158p at the end of it, £1 × 1.158 × 1.083 at the end of the following year and so on.

The overall effect of the five years is an increase by a scale factor of

$$1.158 \times 1.083 \times 1.134 \times 1.180 \times 1.119 = 1.878$$

corresponding to 87.8% over the five years.

A scale factor x per year, applied over the five years, would have the effect of x^5. Thus, for it to be equivalent,

$$x^5 = 1.878$$

$$x = \sqrt[5]{1.878} = 1.134$$

This is the geometric mean of the five figures.

Since 1.134 – 1 = 0.134 or 13.4%, the average inflation rate was 13.4%.

goodness of fit *n* how well a given relationship or distribution fits a particular set of data.

A measure of goodness of fit for observed sample data, suitably grouped, is given by

$$X^2 = \sum_{\text{all groups}} \frac{(f_0 - f_e)^2}{f_e}$$

where f_0 is the observed frequency in a group, and f_e the expected frequency. See CHI-SQUARED TEST, KOLMOGOROV-SMIRNOV TESTS.

grade a position or degree in a scale, as of quality, rank, size, or progression.

Graeco-Latin square see EXPERIMENTAL DESIGN.

graph a drawing depicting the relationship between variables by means of a series of dots, lines, etc., plotted with reference to a set of axes.

Graphs of empirical data

When empirical data are plotted, it is often appropriate to draw a line (or curve) of *best fit* through them (as shown in Fig. 52(a)). The points should not necessarily be joined (Fig. 52(b)).

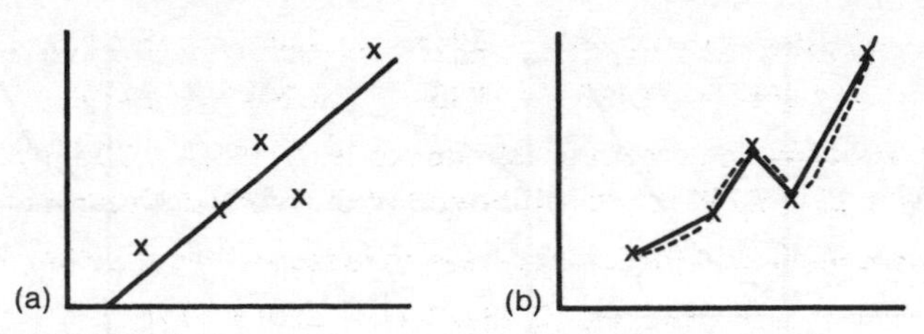

Fig. 52. **Graph.** (a) A line of best fit for a set of data points. (b) Data points in 52(a) possibly incorrectly joined.

Cartesian graphs

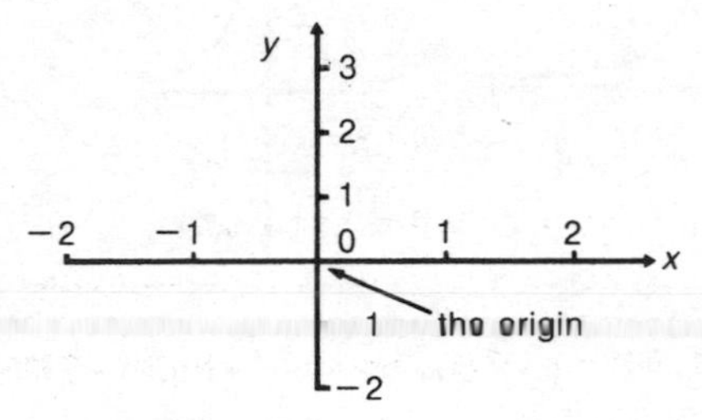

Fig. 53. **Graph.** Axes for a cartesian graph.

The most common type of graph is a *cartesian graph,* with uniform scales on two perpendicular axes, often known as the x- and y- axes (Fig. 53). The x-axis is usually used for the independent variable, the y-axis for the dependent variable. The general equation for the Cartesian graph of a straight line is given by $y = mx + c$ (Fig. 54). This line has gradient m and intercept c (it crosses the y-axis at the point $(0, c)$). Other common Cartesian graphs are shown in Fig. 55.

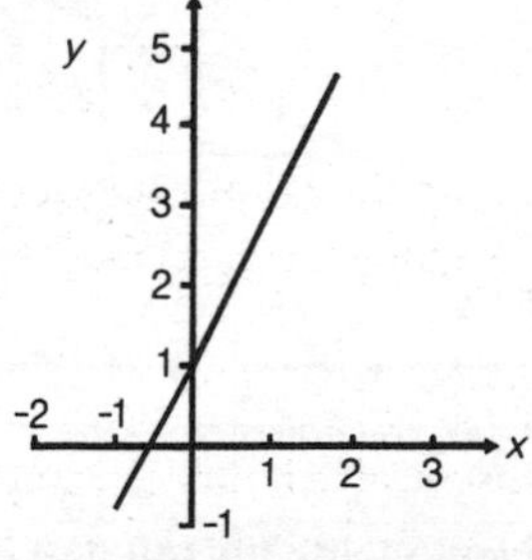

Fig. 54. **Graph.** The straight line, $y = mx+c$, illustated by $y = 2x+1$; the gradient m is 2, the intercept c is 1.

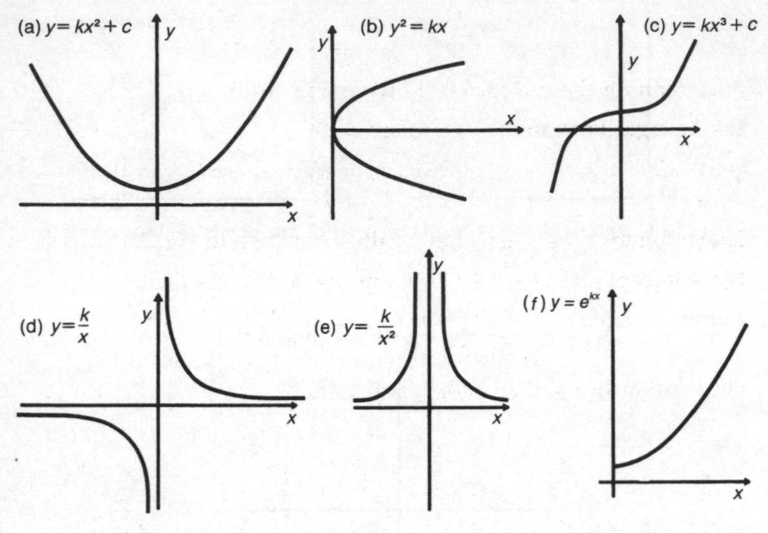

Fig. 55. **Graph.** Some common Cartesian graphs, where k, c are positive constants. (a) $y = kx^2+c$ (b) $y^2 = kx$ (c) $y = kx^3 + c$ (d) $y = \frac{k}{x}$ (e) $y = \frac{k}{x^2}$ (f) $y = e^{kx}$.

Logarithmic graphs. The relationship $y = ax^b$, written in logarithmic form, is:

$$\log y = b \log x + \log a$$

So the graph of $\log y$ against $\log x$ is a straight line with intercept $\log a$ and gradient b. The straight line form of this graph makes it possible, in certain cases, to determine the relationship between two variables from experimental data. (see Fig. 56).

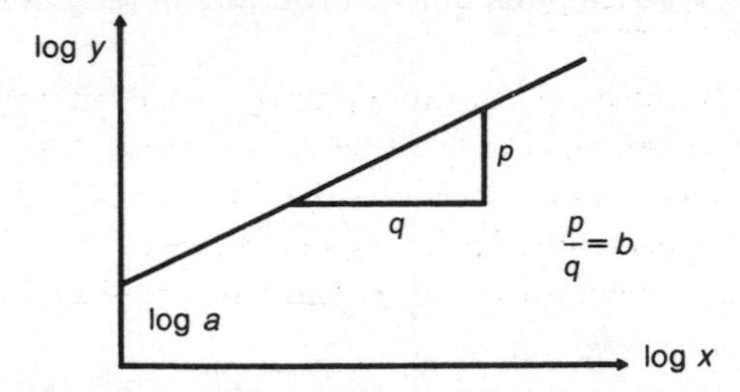

Fig. 56. **Graph.** Logarithmic graph, illustrating the intercept ($\log a$) and the gradient b for $y = ax^b$.

Example: An experiment yields the following pairs of values for the variables x and y.

x	1	2	3	4	5
y	5	17	48	78	125

The graph for these values is shown in Fig. 57(a).

Taking logarithms to the base 10 gives

$\log x$	0	0.301	0.477	0.602	0.699
$\log y$	0.699	1.23	1.68	1.89	2.10

The graph for these logarithmic values is shown in Fig. 57(b).

The intercept gives $\log a = 0.699$; thus $a = 5$. The gradient is:

$$\frac{2.10-0.699}{0.699-0} = 2.0\text{ ; thus } b = 2$$

The relationship plotted is therefore $y = 5x^2$

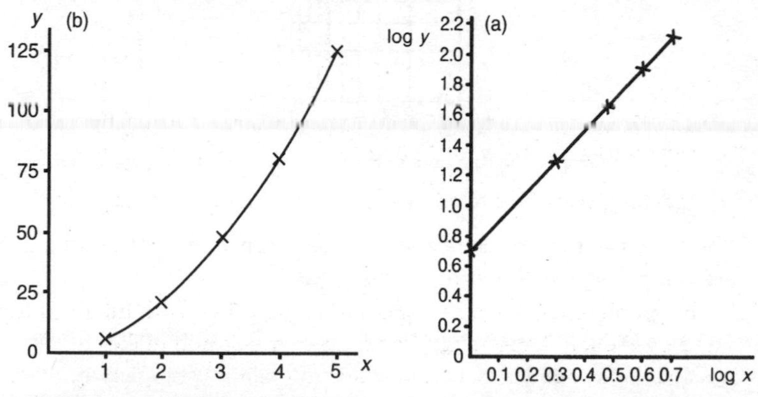

Fig. 57. **Graph** (a) the x-y graph gives a curve. (b) The logarithmic graph gives a straight line. The intercept gives $\log a = 0.699$; thus $a = 5$. The gradient is 2; thus $b = 2$. The relationship plotted is therefore $y = 5x^2$.

The work in producing a logarithmic graph can be reduced by using the alternative method, of plotting the data directly on log-log graph paper, which has non-uniform scales (Fig. 58). Not all data will give a straight line when plotted this way, only those points where the relationship is of the form $y = ax^b$, where a and b are constants.

Semi-logarithmic graphs

These are used for relationships of the form $y = a \times b^x$. In such cases $y = \log a + x \log b$ and so the graph of $\log y$ against x is a straight line with intercept $\log a$ and gradient $\log b$. Semi-logarithmic graph paper has a log scale on one axis (y) and a uniform scale on the other axis (x).

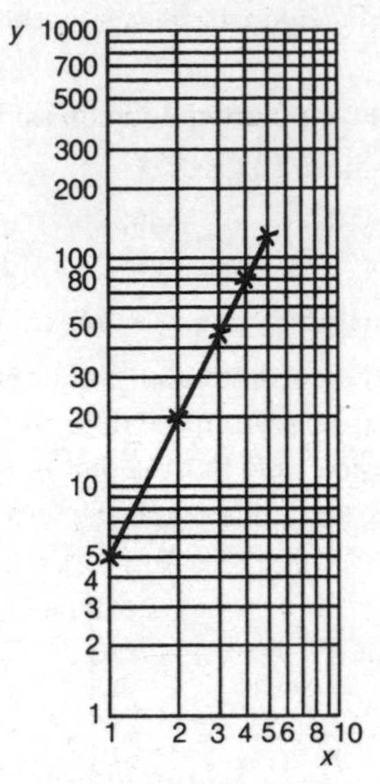

Fig. 58. **Graph.** The data used in Fig. 57 plotted on log-log graph paper.

Square law graphs

These are used when the relationship between x and y (when a and b are constants) is of the form,

$$y = a + bx^2$$

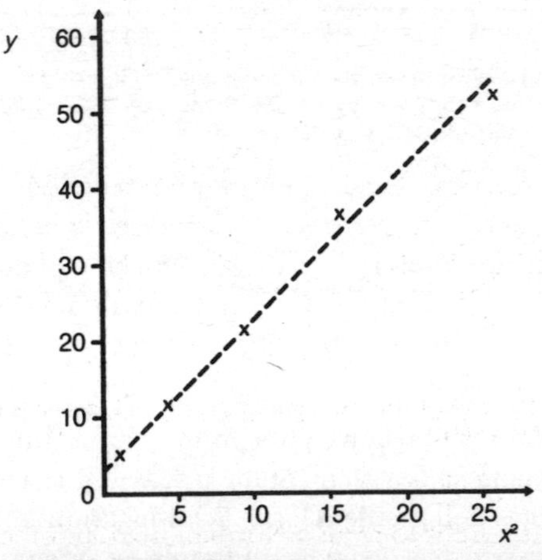

Fig. 59. **Graph.** Data plotted as a square law graph.

In this case, the graph of y against x^2 is a straight line, with intercept a, gradient b.

Example: the data below are plotted as a square law graph in Fig. 58.

x	1	2	3	4	5
y	5	11.2	20.6	35.3	52.1
x^2	1	4	9	16	25

Polar graphs

Polar graphs are centred on a particular point, the pole; the coordinates are the distance, r, from the pole and the direction, θ, between the line joining the pole to the point in question and a fixed direction, usually the x-axis. Polar graph paper may be used.

Example: a simple radio aerial is used to pick up a signal from a distant source. The strength of the signal depends on the orientation of the aerial, as shown in the graph, Fig. 60.

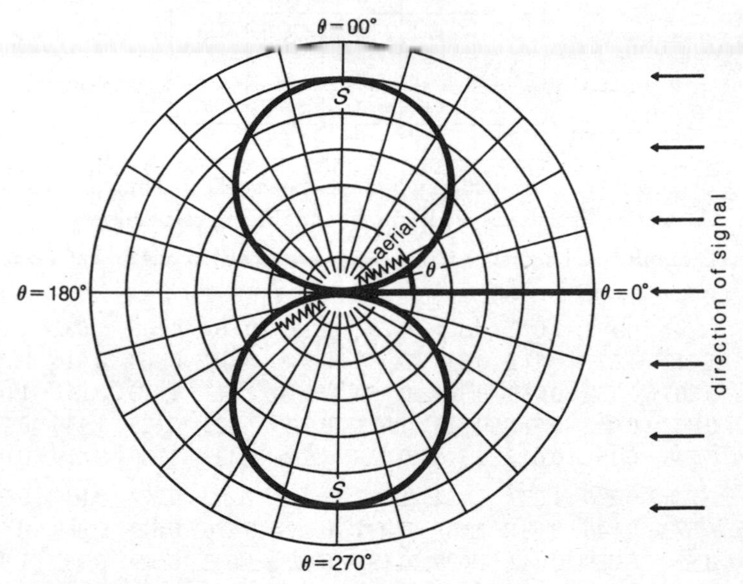

Fig. 60. **Graph.** Polar graph; the strength of the signal, S, depends on the orientation θ, of the aerial.

Normal probability graph paper

Data which are thought to form a Normal distribution can be plotted on Normal probability graph paper.

Several *charts* are published, from which probabilities can be read; they are a form of table. Examples include those used for poisson distribution, binomial distribution, confidence intervals for correlation coefficients, etc.; see also FISHER'S z TRANSFORMATION.

group **1.** a number of people or things considered as a collective unit.

2. a set of elements under an operation, obeying particular axioms (closure, associativity, existence of an identity element and every element having an inverse). Groups are usually associated with abstract algebra but have applications in statistics in experimental design.

The table below illustrates a combination table for a group of four elements under an operation *. The pattern is a Latin square; each element occurs once in each row and each column. See EXPERIMENTAL DESIGN.

*	*a*	*b*	*c*	*d*
a	*a*	*b*	*c*	*d*
b	*b*	*a*	*d*	*c*
c	*c*	*d*	*a*	*b*
d	*d*	*c*	*b*	*a*

grouped data information which has been collected into groups or classes for display, easy reading, or simplifying calculation.

Example: the heights of 100 waves were recorded one day at a coastal research station. They were as follows (in m):

0.701	0.192	1.002	0.824	0.702	0.201	0.889	0.546	1.082	1.103
1.013	1.271	0.212	0.842	0.936	0.836	0.402	0.963	0.614	0.297
0.303	0.757	0.768	0.324	0.959	0.863	1.112	1.173	0.812	1.054
0.867	0.942	0.714	0.723	1.521	0.912	1.223	0.625	1.145	0.161
0.746	0.182	0.646	1.216	0.496	0.934	0.433	1.046	0.419	0.688
0.736	0.296	1.289	1.024	1.362	0.946	1.292	0.679	0.836	0.781
0.573	0.846	0.584	0.637	0.991	0.450	0.476	0.812	0.369	0.849
0.836	0.658	0.742	1.039	0.691	0.823	1.082	0.529	0.962	1.330
0.739	1.154	0.492	0.929	1.234	0.824	0.770	1.341	1.356	0.538
1.256	0.516	0.669	0.550	1.190	0.998	0.199	0.792	0.479	0.901

Looking at these figures as they stand gives very little immediate information to the reader. When, however, they are grouped in classes they are much easier to interpret.

Height (h m)	$0 \le h < 0.2$	$0.2 \le h < 0.4$	$0.4 \le h < 0.6$	$0.6 \le h < 0.8$
Frequency	4	7	15	22
Height (h m)	$0.8 \le h < 1.0$	$1.0 \le h < 1.2$	$1.2 \le h < 1.4$	$1.4 \le h < 1.6$
Frequency	22	14	11	1

The mean $\bar{x}$ and standard deviation of grouped data are estimated by assuming all the members of each group to be at the mid-point of the group.

Height h	*Mid-point* x	*Frequency* f	xf	$(x-\bar{x})$	$(x-\bar{x})^2$	$(x-\bar{x})^2 f$
$0 \le h < 0.2$	0.1	4	0.4	-0.7	0.49	1.96
$0.2 \le h < 0.4$	0.3	7	2.1	-0.5	0.25	1.75
$0.4 \le h < 0.6$	0.5	15	7.5	-0.3	0.09	1.35
$0.6 \le h < 0.8$	0.7	22	15.4	-0.1	0.01	0.22
$0.8 \le h < 1.0$	0.9	26	23.4	0.1	0.01	0.26
$1.0 \le h < 1.2$	1.1	14	15.4	0.3	0.09	1.26
$1.2 \le h < 1.4$	1.3	11	14.3	0.5	0.25	2.75
$1.4 \le h < 1.6$	1.5	1	1.5	0.7	0.49	0.49
Total		$n = 100$				10.04

$$\textit{Mean, } \bar{x} = \frac{\sum xf}{\sum n} = \frac{80}{100} = 0.8 \text{ (estimated)}$$

$$\textit{Variance, } s^2 = \frac{\sum (x-\bar{x})^2 f}{n-1} = \frac{10.04}{99} = 0.101 \text{ (3 s.f.)}$$

$$\textit{Standard deviation, } s = \sqrt{\text{Variance}} = 0.318 \text{ (3 s.f.)}$$

If the distribution has a central mode and tails off to either side, as in the example above, this procedure will involve errors.

For groups below the mode, more of the members of a group are above the mid-point than below it (Fig. 61). For groups above the mode, the opposite is true.

For estimation of the mean, the errors above and below will approximately cancel each other out, so the method described is reasonably accurate. For variance and standard deviation, however, the errors will add up so that the estimated value will be too large (some of the population is nearer the middle than estimated).

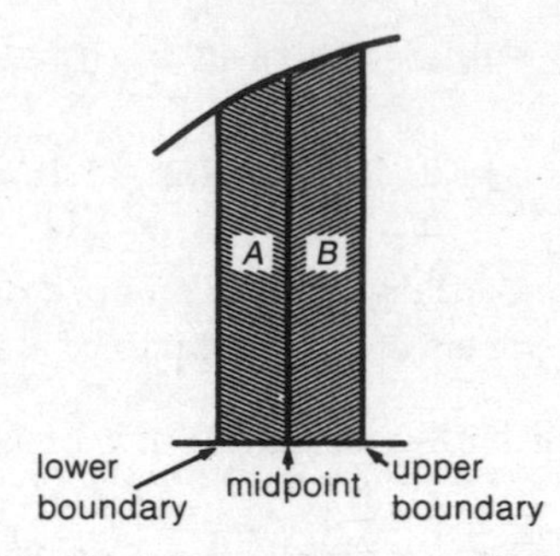

Fig. 61. **Grouped data.** The frequency above the midpoint of this class (region *B*) is greater than that below the midpoint (region *A*). This is often the case for a class below the mode.

This can be allowed for by applying *Sheppard's correction*:

$$\textit{Corrected variance} = \textit{Calculated variance} - \frac{c^2}{12}$$

where c is the class interval.

For the figures in the example of wave heights:

$$\textit{Corrected variance} = 0.101 - \frac{(0.2)^2}{12} = 0.0981 \text{ (3 s.f.)}$$

$$\textit{Corrected standard deviation} = 0.313\text{m (3 s.f.)}$$

Sheppard's correction is only of value if the distribution is approximately Normal. Thus it must be continuous (or discrete with no gaps in the data), and must tail off to zero either side of the mode. In addition, the class interval must be less than the standard deviation. Sheppard's correction is only occasionally used.

h

harmonic mean the harmonic mean m of the n numbers $x_1, x_2, \ldots x_n$ is given by:

$$\frac{1}{m}=\frac{1}{n}\left\{\frac{1}{x_1}+\frac{1}{x_2}+\ldots+\frac{1}{x_n}\right\} \text{ or by } m=\frac{n}{\sum_{i=1}^{n}\frac{1}{x_i}}$$

Thus, the harmonic mean of 2, 4 and 5 is calculated as follows:

$$\frac{1}{m}=\frac{1}{3}\left(\frac{1}{2}+\frac{1}{4}+\frac{1}{5}\right)$$

giving

$$\frac{1}{m}=\frac{19}{60} \text{ and so } m=\frac{60}{19}=3\frac{3}{19}$$

Example: a man travels 1 km at 2 km h^{-1}, 1 km at 4 km h^{-1} and 1 km at 5 km h^{-1}. What is his average speed?

Thus, the average speed is the harmonic mean of the three speeds.

$$\textit{Average speed}=\frac{\text{Distance travelled}}{\text{Time taken}}$$

$$=\frac{3\text{km}}{\left(\frac{1}{2}+\frac{1}{4}+\frac{1}{5}\right)\text{h}}=3\tfrac{3}{19}\text{ km h}^{-1}$$

heteroscedastic **1.** (of a number of distributions) having different variances.

2. (of a bivariate or multivariate distribution) not having any variable with constant variance for all values of the other or others.

3. (of a random variable in a multivariate distribution) having different variances for different values of the other variables.

Compare HOMOSCEDASTIC. See also BIVARIATE DISTRIBUTION.

hierarchy an organisation of objects, or groups of objects, in a graded order. See CLUSTER ANALYSIS.

hinge see QUARTILE.

histogram a chart for displaying grouped continuous data in which the width of each bar is proportional to the class interval, and the area of each bar is proportional to the frequency it represents.

Example: In a survey of a village, 126 householders were asked how long they had lived there. The results were grouped as follows.

Duration of residence, d (years)	$0 \leq d < 1$	$1 \leq d < 5$	$5 \leq d < 10$	$10 \leq d < 20$	$20 \leq d < 50$
Frequency (number of householders)	6	39	30	27	24

The histogram displaying this information is shown in Fig. 62.

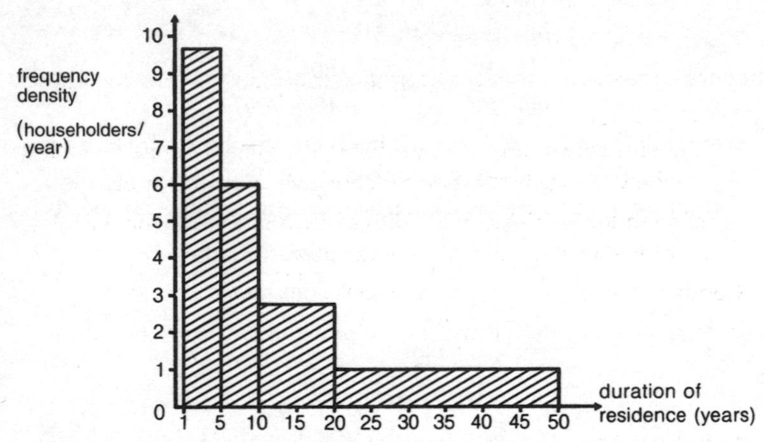

Fig. 62. **Histogram.** Results of a survey of householders in a village of how long they had lived there.

The heights of the bars are calculated by dividing the frequencies by the class intervals, measured in suitable units, in this case years.

Class	*Frequency*	*Class interval*	*Frequency density Height of bar*
$0 \leq d < 1$	6	1	6
$1 \leq d < 5$	39	4	9.75
$5 \leq d < 10$	30	5	6
$10 \leq d < 20$	27	10	2.7
$20 \leq d < 50$	24	30	0.8

Notice that the vertical scale is "Frequency density" and *not* "Frequency". In this case Frequency density is calculated as frequency per year. Notice also that it is the areas (not the heights) of the bars which give the frequencies. In this case:

$$(1\text{–}0) \times 6 = 6$$
$$(5\text{–}1) \times 9.75 = 39$$
$$(10\text{–}5) \times 6 = 30$$
$$(20\text{–}10) \times 2.7 = 27$$
$$(50\text{–}20) \times 0.8 = 24$$

This is in contrast to a bar chart, a frequency chart or a vertical line chart where the heights do represent frequencies.

The term histogram is, however, sometimes applied loosely (and incorrectly) to diagrams with equal class intervals in which the vertical scale is frequency.

historigram see TIME SERIES.

homoscedastic **1.** (of a number of distributions) having the same variance.

2. (of a bivariate or multivariate distribution) having one variable of which the variance is constant for all values of the other or others.

3. (of a random variable in a multivariate distribution) having constant variance for all values of the other variables.

Compare HETEROSCEDASTIC. See also BIVARIATE DISTRIBUTION.

hypergeometric distribution the probability distribution p(x) given by:

$$\mathrm{p}(x) = \frac{\binom{a}{x}\binom{b}{n-x}}{\binom{a+b}{n}}$$

where $0 \le x \le a$; $n \le a+b$; a and b are positive integers, and $\binom{a}{x}$ etc. are binomial coefficients.

For this distribution,

$$Mean = \frac{na}{a+b}$$

$$Variance = \frac{nab(a+b-n)}{(a+b)^2(a+b-1)}$$

This distribution is generated when a sample of size n is taken (without replacement) from a population containing a items of one particular type, and b of others. The probability that the sample contains exactly x of the particular type is given by the relevant term in the hypergeometric distribution.

Example: In a pile of 20 suspect banknotes, 12 are forged and the other 8 genuine. A random sample of 5 of the notes is taken. What is the probability that exactly 2 of them are forged?

In this case $a = 12$, $b = 8$, $n = 5$ and $x = 2$, so the required probability is given by substituting these values into the equation for p(x).

$$\begin{aligned} p(2) &= \frac{\binom{12}{2}\binom{8}{3}}{\binom{20}{5}} \\ &= \frac{66 \times 56}{15504} \\ &= 0.238 \end{aligned}$$

See BINOMIAL COEFFICIENTS.

hypothesis a theory which is put forward either because it is believed to be true or because it is to be used as a basis for argument, but which has not been proved. Statistical hypotheses often refer to the values of parameters, like mean and variance, of the parent population from which a sample has been drawn. Setting up and testing such hypotheses forms part of inferential statistics.

See NULL HYPOTHESIS.

i

impossible incapable of occurring or happening. If an event is impossible, the probability of its occurrence is zero.

increment **1.** a small change in the value of a variable.

2. to change the value of a variable systematically by small amounts.

independent events events which have no influence on each other. If a die and a coin are thrown together, the events 'The die shows 6' and 'The coin comes heads' are independent.

If the knowledge that an event *A* has occurred influences the probability of another event *B* occurring, then *B* is not independent of *A*. For example, the event "The temperature on a given day will exceed 20°C" is unlikely to be independent of the event "The day selected is in August".

This is expressed mathematically by saying that event A is independent of event B if

$$P(A|B) = P(A|B') = P(A)$$

In the same way, event B is independent of event A if

$$P(B|A) = P(B|A') = P(B)$$

It is not difficult to prove that if *A* is independent of *B*, then *B* must be independent of *A* and

$$P(A \cap B) = P(A)P(B)$$

See also CONDITIONAL PROBABILITY.

independent random variables random variables, e.g. *X* and *Y*, such that a knowledge of the value of X does not affect the probability distribution of *Y*, and vice versa. Thus there is no relationship between the values taken by independent random variables; their covariance is zero.

When two dice are thrown, their scores are independent. By contrast, the marks for papers 1 and 2 of a mathematics examination are not independent, with many candidates performing well, moderately, or badly on both papers.

Thus for independent discrete random variables X and Y

$$P(X = x \text{ and } Y = y) = P(X = x) \times P(Y = y)$$

for all values of x and y. See RANDOM VARIABLE.

independent variable see DEPENDENT VARIABLE, EXPLANATORY VARIABLE.

index number a statistic giving the value of a quantity (like the crime rate, or the cost of living) relative to its level at some fixed time or place which, conventionally, is given the number 100.

Example: In a certain country, the annual numbers of murders during the years 2005-2009 were as follows:

2005	2006	2007	2008	2009
500	515	520	535	485

If 2005 is taken as the base year, and assigned the index number, 100, then the index number for 2006 is calculated as:

$$\frac{515}{500} \times 100 = 103$$

The full set of index numbers is thus

2005	2006	2007	2008	2009
100	103	104	107	97

In this example, the index was easily calculated because only one item, the number of murders, was under consideration. Index numbers which are used in practice, like the Index of Retail Prices, are usually built up from many items, and are calculated as a weighted mean of the indices for each of the items.

Example: a very simple cost-of-living index is constructed from three items, petrol, meat, and potatoes, in the ratio 3:5:2. In the year taken as the base, and in the current year, prices were:

	Base year	*Current year*
Petrol (1 l)	£1	£1.70
Meat (250 g)	£2.50	£2.80
Potatoes (10 kg)	£1.50	£1.20

The individual index numbers are calculated as follows:

The Base Year indices are all 100; the Current Year indices are

$$\text{Petrol} \quad \frac{1.70}{1} \times 100 = 170$$

$$\text{Meat} \quad \frac{2.80}{2.50} \times 100 = 112$$

$$\text{Potatoes} \quad \frac{1.20}{1.50} \times 100 = 80$$

The calculation of the cost-of-living index is then continued in the table below:

	Base index	Current index number		Weighting		Current cost of living
Petrol	100	170	×	3	=	510
Meat	100	112	×	5	=	560
Potatoes	100	80	×	2	=	160
		Total		10		1,230

$$\text{Current index number} = \frac{1230}{10} = 123$$

The question of the best weighting to give the different items in constructing an index number is one which has no correct answer. In the case of an actual cost of living index, there might realistically be 600 component items. A correct weighting for one person need not be so for another; for example, for one man, beer may be a major source of expenditure while another, being teetotal, spends nothing at all on it. In practice, a weighting scheme is determined on the basis of the average amount spent on each item. However, as tastes change with time, so does the ideal weighting; furthermore, as new items become available and old ones obsolete, the list itself is liable to change. In the UK, the official index of prices is re-based from time to time because of this, as (in another context) are stock market indices.

In *Laspeyre's Index* the weighting is taken from the base year, in *Paasche's Index* from the current year. This means that Laspeyre's Index tends to emphasize items which might be out of date and to ignore changes in taste or lifestyle over the years. Paasche's Index, on the other hand, runs into difficulty with new items which are significant in the current year but have no base value because they did not exist or were not used before. Imagine, for example, a cost of living index, set up with base year 1800 to compare the cost of living then with now.

If Laspeyre's index is used, items like candles and horse feed will be quite heavily weighted, whereas electricity and petrol would not feature at all. If Paasche' index is used, the reverse is the case.

A compromise between the two is the *Typical Year Index*, where the weightings are taken for some typical time between the base year and the present.

In *Fisher's Ideal Index*, the value is taken as the geometric mean of Laspeyre's index and Paasche's index. The *Marshall Edgeworth Bowley Index* takes the weighting for any item to be the average of that for the base year and that for the current year.

Formulae for the various indexes are as follows, where p_o = base year price, p_n = current year price, q_o = base year weighting, q_n = current year weighting and q_t = typical year weighting:

$$\textit{Laspeyre's Index} = \frac{\sum p_n q_0}{\sum p_0 q_0}$$

$$\textit{Paasche's Index} = \frac{\sum p_n q_n}{\sum p_0 q_n}$$

$$\textit{Typical Year Index} = \frac{\sum p_n q_t}{\sum p_0 q_t}$$

$$\textit{Fisher's Ideal Index} = \sqrt{\left\{\frac{\sum p_n q_0}{\sum p_0 q_0}\right\}\left\{\frac{\sum p_n q_n}{\sum p_0 q_n}\right\}}$$

$$\textit{Marshall Edgeworth Bowley Index} = \frac{\sum p_n (q_0 + q_n)}{\sum p_0 (q_0 + q_n)}$$

inference (as used in statistics) drawing conclusions about a parent population on the basis of evidence obtained from a sample.

inferential statistics that branch of statistics which involves drawing conclusions about a population from sample data. This often involves hypothesis testing. Compare DESCRIPTIVE STATISTICS. See also INTERPRETIVE STATISTICS.

inferior see SUPERIOR.

infinite having no limits or boundaries (of time, space, extent or magnitude).

infinitesimal **1.** infinitely or immeasurably small.

2. of, relating to, or involving a small change in the value of a variable.

inlier a term sometimes used for a data value that is believed to have come from a different parent population but which lies well within the main body of a set of sample data, making it difficult or impossible to detect as an error by statistical analysis. Compare OUTLIER.

integer a whole number; 5, -17 and 0 are all integers.

intelligence quotient, IQ a measure of the intelligence of an individual. When first introduced, IQ was designed to have mean value 100, standard deviation 15. Since then, measurements based on standard tests show an increasing trend in the mean.

interaction a term describing the relationship between two (or more) factors, say A and B, whose levels affect the outcome of an experiment or process when, in addition, the effect of A on the outcome depends on the level of B, and that of B on the level of A.

Example: a company is developing two new fertilisers, A and B, for use on crops. There is just one form of fertiliser A but there are three forms of B, denoted by B_1, B_2 and B_3. Experiments are carried out using A and each of the three forms of B at both Low and High levels. The resulting yields (in suitable units) are summarised in the tables below.

Case 1: A and B_1

		Level of B_1	
		Low	High
Level of A	Low	3	7
	High	9	13

There is no interaction between A and B_1. Moving from Low A to High A increases the yield by 6. Moving from Low B_1 to High B_1 increases the yield by 4. Thus the A effect is the same at both levels of B_1 and the B_1 effect is the same at both levels of A.

Case 2: A and B_2

		Level of B_2	
		Low	High
Level of A	Low	3	7
	High	9	16

There is positive interaction between A and B_2. The yield for High A/High B_2 is 3 greater than it would be just from the effects of A and B_2 alone.

Case 3: A and B_3

		Level of B_3	
		Low	High
Level of A	Low	3	7
	High	9	10

There is negative interaction between A and B_3. The yield for High A/High B_3 is 3 less than it would be just from the effects of A and B_3 alone.

Interaction is an important concept in statistics. See FACTORIAL EXPERIMENT.

interpolation estimation of a value of a variable between two known values. Compare EXTRAPOLATION.

Example: Alexandra was 1.21 m tall on 1 January 2002 and 1.25 m tall a year later, on 1 January 2003. Estimate her height on 1 April 2002.

This could be done graphically or by calculation.

1 Jan to 1 April = 3 months
One year = 12 months

So her height on 1 April was:

$$\tfrac{3}{12}\times(1.25-1.21)+1.21=1.22 \text{ m.}$$

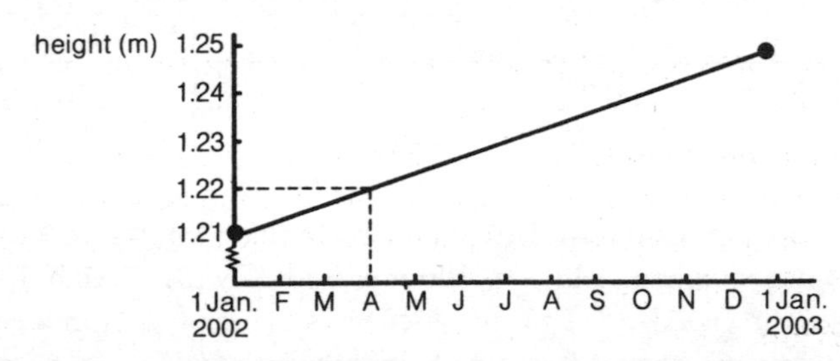

Fig. 63. **Interpolation.** From the actual values plotted for Alexandra's height on 1 January 2002 and 1 January 2003, her height on 1 April 2002 can be interpolated (1.22 m).

From the graph (Fig. 63) it can be estimated by interpolation that Alexandra's height on 1 April was 1.22 m. This answer assumes that the graph of Alexandra's height plotted against time was a straight line, indicated that she grew uniformly over the year. This may very well have not been the case; she could for example, have grown more in the second half of the year than in the first, or vice versa. However, in the absence of other information, it is usual to assume uniform change; the method is then called *linear interpolation*.

It is common to use linear interpolation when reading from tables.

Example: These figures come from the table for the height of the Normal curve (see TABLE 1):

x	1.0	1.1
$\varphi(x)$	0.242	0.218

An estimate for the value of $\varphi(1.06)$, found by linear interpolation, is given by:

$$\varphi(1.06) = 0.242 + \frac{(1.06-1.00)}{(1.10-1.00)} \times (0.218-0.242) = 0.2276$$

Non-linear interpolation is sometimes used when enough points are known to draw a curved graph as, for example, when finding intermediate values from a cumulative frequency curve.

interpretive statistics that branch of statistics concerned with drawing conclusions from data. Compare DESCRIPTIVE STATISTICS.

interquartile range the difference between the upper and lower quartiles of a set of numbers. For example, the numbers

3 4 4 4 6 6 6 6 7 7 8 9

have upper quartile 7, lower quartile 4; the interquartile range is thus 7-4 = 3

Half of the interquartile range (in this example, $\frac{1}{2} \times 3 = 1\frac{1}{2}$) is called the *semi-interquartile range* (or *quartile deviation*) and is a measure of dispersion or spread comparable with standard deviation and mean absolute deviation. See QUARTILE.

intersection see SET.

interval estimation see POINT ESTIMATION.

interval scale a scale in which equal differences between pairs of points anywhere on the scale are represented by equal intervals, but the zero point is arbitrary. The time interval between the starts of years 2006 and 2016 is the same as that between the starts of 1066 and 1076, namely 10 years. The zero point, year 1 AD, is arbitrary; time did not begin then. Other examples of interval scales include the heights of tides, and the measurement of longitude.

inverse Fisher transformation see FISHER'S *z* TRANSFORMATION.

irregular variation that variation within a time series which is not accounted for by trend, seasonal components or cyclic components. See TIME SERIES.

k

Kendall's coefficient of concordance (of largely historical interest) a measure of the level of agreement (concordance) between several sets of ranked data. It takes values between 0 and 1. Symbol: W.

Those doing the ranking are described as "judges". If there are only two judges, a value of W near 0 means disagreement between them. If there are many judges a value of 0 means randomness in their judgements. A value near 1 shows agreement between the judges. See also CORRELATION.

Kendall's coefficient of concordance, W, is calculated using the formula:

$$W = \frac{12S}{K^2(N^3 - N)}$$

where $S = \sum_{i=1}^{N} \left(R_i - \bar{R}\right)^2$ (the example below shows how this is worked out), K is the number of judges, and N is the number of objects judged.

Example: in testing a new range of cakes, labelled C_1 to C_6, a bakery asks four people to rank them in order of preference.

The results and the subsequent calculation of W are as follows.

	C_1	C_2	C_3	C_4	C_5	C_6
Mr McDiarmid	1	6	2	5	3	4
Mrs Nandi	2	5	1	3	4	6
Ms O'Leary	1	4	3	6	5	2
Mr Patrice	2	4	3	6	1	5
Sum of ranks R_i	6	19	9	20	13	17

Total sum of ranks = 6 + 19 + 9 + 20 + 13 + 17 = 84

Mean of ranks $\bar{R}$ $\frac{84}{6} = 14$

	C_1	C_2	C_3	C_4	C_5	C_6
$(R_i - \bar{R})$	-8	5	-5	6	-1	3
$(R_i - \bar{R})^2$	64	25	25	36	1	9

Sum $S = 64 + 25 + 25 + 36 + 1 + 9 = 160$

$$W = \frac{12 \times 160}{16(216 - 6)} = 0.57$$

Kendall's coefficient of rank correlation see CORRELATION COEFFICIENT.

Kolmogorov-Smirnov tests a set of tests for goodness of fit based on comparing the distribution function of a sample with the theoretical distribution function of a proposed model, or (in the case of the 2-sample test) with that of another sample.

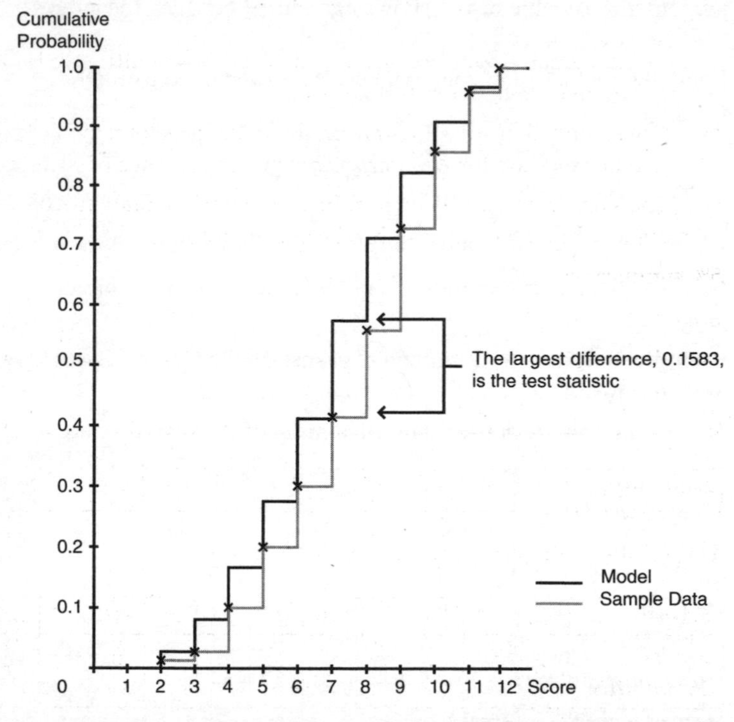

Fig. 64. **Kolmogorov-Smirnov tests.** Goodness of fit test (example 1) for a completely specified distribution.

In all the tests the cumulative distribution of the sample data is compared with that for the distribution with which it is being compared. The largest difference, *D*, between the two cumulative distributions is then found.

The different tests are illustrated in the following examples.

1. *Kolmogorov-Smirnov goodness of fit test for completely specified distributions*

Example: it is suggested that a particular gambler is able to influence the outcome by the way he throws dice. Unknown to him, a record is kept of the outcomes when he throws a pair of dice 80 times one evening.

Score	2	3	4	5	6	7	8	9	10	11	12
Frequency	1	1	6	8	8	10	11	14	10	8	3

Use the Kolmogorov-Smirnov test to test at the 5% significance level whether there is any evidence of cheating

H_0: The sample data are drawn from the population obtained when two fair dice are thrown fairly. There is no evidence of cheating.

H_1: The sample data are drawn from a different population. There is evidence that the gambler has been cheating.

5% significance level

2-tail test

The probability distribution of the total scores when two fair dice are thrown fairly is

Score	2	3	4	5	6	7	8	9	10	11	12
Probability	$\frac{1}{36}$	$\frac{2}{36}$	$\frac{3}{36}$	$\frac{4}{36}$	$\frac{5}{36}$	$\frac{6}{36}$	$\frac{5}{36}$	$\frac{4}{36}$	$\frac{3}{36}$	$\frac{2}{36}$	$\frac{1}{36}$

This can be written as a cumulative distribution.

Score	≤2	≤3	≤4	≤5	≤6	≤7	≤8	≤9	≤10	≤11	≤12
Model Probability	$\frac{1}{36}$ 0.0278	$\frac{3}{36}$ 0.0833	$\frac{6}{36}$ 0.1667	$\frac{10}{36}$ 0.2778	$\frac{15}{36}$ 0.4167	$\frac{21}{36}$ 0.5833	$\frac{26}{36}$ 0.7222	$\frac{30}{36}$ 0.8333	$\frac{33}{36}$ 0.9167	$\frac{35}{36}$ 0.9722	$\frac{36}{36}$ 1

The cumulative probability distribution for the sample data is

Score	≤2	≤3	≤4	≤5	≤6	≤7	≤8	≤9	≤10	≤11	≤12
Data Probability	$\frac{1}{80}$ 0.0125	$\frac{2}{80}$ 0.025	$\frac{8}{80}$ 0.1	$\frac{16}{80}$ 0.2	$\frac{24}{80}$ 0.3	$\frac{34}{80}$ 0.425	$\frac{45}{80}$ 0.5625	$\frac{59}{80}$ 0.7375	$\frac{69}{80}$ 0.8625	$\frac{77}{80}$ 0.9625	$\frac{80}{80}$ 1

The two cumulative probability distributions are illustrated in Fig. 64. The difference between the cumulative probabilities for the two distributions are then found.

Score	≤2	≤3	≤4	≤5	≤6	≤7	≤8	≤9	≤10	≤11	≤12
Model	0.0278	0.0833	0.1667	0.2778	0.4167	0.5833	0.7222	0.8333	0.9167	0.9722	1
Data	0.0125	0.025	0.1	0.2	0.3	0.425	0.5625	0.7375	0.8625	0.9625	1
Difference	0.0153	0.0583	0.0667	0.0778	0.1167	0.1583	0.1597	0.0958	0.0542	0.0097	0

In this case the sample size was 80 and so the largest difference is denoted by D_{80}. This is the test statistic.

$$D_{80} = 0.1597$$

Table 14 gives critical values for D_n. For a 2-tail test at the 5% significance level, the critical value for D_{80} is 0.1496.

Since 0.1597>0.1496, the null hypothesis is rejected. At this level, the evidence supports the idea that the gambler is cheating.

2. *Kolmogorov-Smirnov test for Normality*

There are two situations where a Kolmogorov test for Normality may be required. In one situation the mean and variance of the proposed Normal distribution are already known and so it is appropriate to use the Kolmogorov-Smirnov goodness of fit test for completely specified distributions. If, however, the mean and variance are not known and have to be estimated from the sample data, the procedure shown in the next example should be followed. In both cases, the cumulative frequency graph in Table 2 may be found helpful.

Example: a geologist finds 10 rocks which she believes to be meteorites. Their masses in grams are: 34.5, 38.6, 42.5, 42.9, 47.8, 50.2, 53.3, 54.1, 56.0, 60.1. She uses the Kolmogorov-Smirnov test to judge whether these figures could have could have come from a Normal distribution.

H_0: The underlying population is Normal.

H_1: The underlying population is not Normal.

5% significance level

2-tail test

The sample data are used to estimate the population mean and standard deviation.

Estimated population mean: $\overline{x} = \dfrac{480}{10} = 48.0$

Estimated population standard deviation: $s = \sqrt{\dfrac{1}{n-1} S_{xx}} = \sqrt{\dfrac{607.46}{9}} =$ 8.22 (to 2 decimal places)

These values are used to find z-values for the data items, using the transformation $z = \dfrac{x - \overline{x}}{s}$. The corresponding values of $\Phi(z)$, the cumulative Normal distribution, are then found using Table 15.

The differences between the values of $\Phi(z)$ and the corresponding cumulative probabilities of the items within the data set are then calculated.

Value	34.5	38.6	42.5	42.9	47.8	50.2	53.3	54.1	56.0	60.1
z-value	-1.642	-1.144	-0.669	-0.620	-0.024	0.268	0.645	0.742	0.973	1.472
$\Phi(z)$	0.0503	0.1263	0.2517	0.2676	0.4904	0.6057	0.7405	0.7710	0.8348	0.9295
Cum pr	0.1	0.2	0.3	0.4	0.5	0.6	0.7	0.8	0.9	1.0
Diff	-0.0497	-0.0737	-0.0483	-0.1324	-0.0096	0.0057	0.0405	-0.0290	-0.0652	-0.0705

The test statistic is the largest absolute difference, in this case 0.1324. Critical values for this test are found in Table 15. The critical value for $n = 10$ for a 2-tail test at the 5% significance level is 0.2619.

Since $0.1324 < 0.2619$, there is no reason to reject the null hypothesis.

3. *Kolmogorov-Smirnov 2-sample test*

In this test the cumulative distributions of two samples are compared.

Example: in an experiment, two diets A and B are tested over the same period of time. The weight losses, in kilograms, of those who complete the experiment are as follows:

A: 4, 4, 6, 7, 7, 8, 8, 10

B: 3, 4, 5, 5, 5, 5, 6, 6, 7, 11

The Kolmogorov-Smirnov 2-sample test is used to decide, at the 2% significance level, whether there are differences in the effects of the two diets.

H_0: The two samples are drawn from the same underlying population. There is no difference in the effects of the two diets.

H_1: The two samples are drawn from different populations.

2% significance level

2-tail test

The cumulative probabilities for the two samples are given in the table below. The last row gives the differences between them.

Wt loss	2	3	4	5	6	7	8	9	10	11
A	0	0	0.25	0.25	0.375	0.625	0.875	0.875	1	1
B	0	0.1	0.2	0.6	0.8	0.9	0.9	0.9	0.9	1
Diff	0	-0.1	0.05	-0.35	-0.425	-0.275	-0.025	-0.025	0.1	0

The greatest absolute value of the difference is 0.425 and this is denoted by D. The test statistic, D^*, is given by $D^* = n_1 n_2 D$ where n_1 and n_2 are the sample sizes.

Thus, in this case, $D^* = 8 \times 10 \times 0.425 = 34$.

The critical values for this test are found in Table 16. For $n_1 = 8$ and $n_2 = 10$ and a 2-tail test at the 2% significance level the critical value is 56.

Since $34 < 56$, the null hypothesis is accepted. At this level the evidence does not support the claim that there is a difference in the two diets.

Although the Kolmogorov-Smirnov two-sample test can be used as a 1-tail test of a difference in a particular direction, it is more usual to use a Mann-Whitney or Wilcoxon test in that situation.

Kruskal-Wallis one-way analysis of variance a non-parametric test of the null hypothesis that three or more samples are drawn from the same population. Symbol of the test statistic: H. The test is carried out on the overall ranks of the items sampled, and so is a test for differences in location of the particular samples. It is, by its nature, a 1-tail test.

Example: 21 expectant mothers followed three different diets, D_1, D_2 and D_3. At birth, the weights of their babies (in kg) were as follows:

D_1	D_2	D_3
2.8	3.4	2.9
3.5	3.2	3.9
2.7	3.3	3.1

D_1	D_2	D_3
3.7	3.3	3.9
3.7	3.6	4.0
3.8	3.0	4.1
2.8		4.2
		4.0

Can it be claimed, at the 5% significance level, that the different diets had an effect on the weight of the babies?

H_0: There is no difference between birth-weights of babies whose mothers have been on the different diets.

H_1: There is a difference in birth-weights, according to diet.

5% significance level

1-tail test

The babies are ranked 1 to 21, according to weight, and then the test statistic H is calculated, where

$$H = \frac{12}{N(N+1)} \left(\sum \frac{R_j^2}{N_j} \right) - 3(N+1)$$

In this case there are three groups, so j takes the values 1, 2, and 3.

N_1, N_2, N_3 are the sizes of the three different groups, totalling N; $N_1 = 7$, $N_2 = 6$, $N_3 = 8$; N is thus $7 + 6 + 8 = 21$.

R_1, R_2, R_3 are the sums of the overall ranks for each group, and are calculated from the table of ranks for each baby, below.

	D_1	D_2	D_3
	19.5	12	18
	11	15	5.5
	21	13.5	16
	8.5	13.5	5.5
	8.5	10	3.5
	7	17	2
	19.5		1
			3.5
Sum of ranks, R_j	95	81	55
R_j^2	9025	6561	3025

$$H = \frac{12}{21 \times (21+1)} \left(\frac{9025}{7} + \frac{6561}{6} + \frac{3025}{8} \right) - 3(21+1) = 5.71$$

The critical value for H is obtained from the χ^2 tables for $K-1$ degrees of freedom, where K is the number of groups. In this example, at the 5% level for 2 degrees of freedom the critical value (from Table 4) is 5.99 (see CHI-SQUARED DISTRIBUTION).

Since $5.71 < 5.99$, the null hypothesis cannot be rejected at this level. Thus, there is no evidence, at the 5% level, on which to claim differences in the weights of babies whose mothers used the three diets. It should, however, be noted that the value of H is close to the critical value. In such situations it is good practice to carry out more tests to clarify the position.

kurtosis the sharpness of a peak on a graph representing a distribution (see Fig. 65). A distribution with high kurtosis is called *leptokurtic*; one with medium kurtosis, *mesokurtic*, and one with low kurtosis, *platykurtic*.

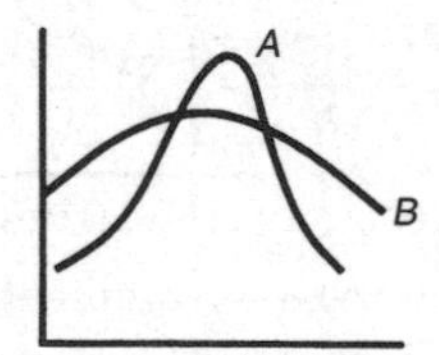

Fig. 65. **Kurtosis.** Distribution *A* has larger kurtosis than *B*.

Population kurtosis is measured by the *coefficient of kurtosis*, which is defined as:

$$\frac{\mathrm{E}[(X-\mu)^4]}{\sigma^4} - 3$$

See EXPECTATION. See also MOMENT.

L

lag see AUTOCORRELATION.

Laspeyre's index see INDEX NUMBER.

latin square see EXPERIMENTAL DESIGN.

lattice diagram in quality control, a diagram on which the results of ATTRIBUTE TESTS on samples are illustrated.

Example: the first six items to be tested are Good, Good, Defective, Good, Defective, Good. Starting at the bottom left of a lattice diagram (Fig. 66), a line is drawn horizontally to the right every time an item is good, vertically up if it is defective. The bent line which results is called the *sampling line.*

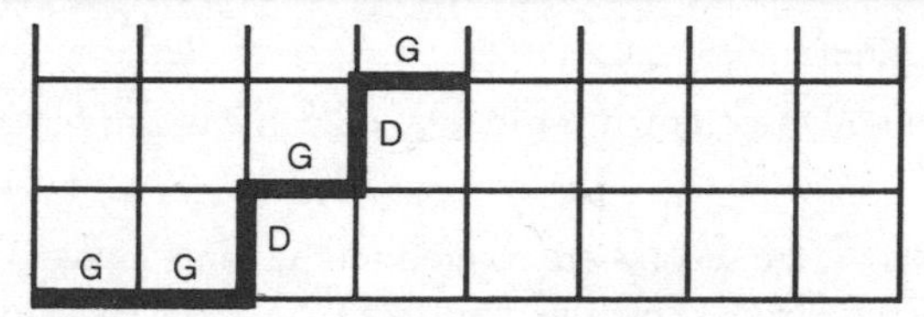

Fig. 66. **Lattice diagram.** This sampling line illustrates whether the first six items to be tested were good G or defective D in a quality-control test.

In a DOUBLE SAMPLING SCHEME, one sample is taken; if the number of defective items is not more than the ACCEPTANCE NUMBER, A_1, the whole batch is accepted. If it is greater than another larger number, A_2, the batch is rejected or subjected to a 100% inspection. If the number of defectives lies between A_1 and A_2, a second sample is taken. If, after the second sample is taken, the total number of defectives is greater than a new acceptance number, A_3, the whole batch is rejected or subjected to a 100% inspection.

Example: First sample: Size = 40 $\quad A_1 = 3 \quad A_2 = 7$

By the end of the first sample, the batch will have been rejected if the number of defectives is over 7, indicated by the upper line on the lattice in Fig. 67(a). It will have been accepted if the outcome is one of:

(40 G, 0 D), (39 G, 1 D), (38 G, 2 D), (37 G, 3 D).

These points form the lower line on the lattice diagram in Fig. 67(a). Thus, these two lines mark the outcomes: *acceptance* (bottom right) and *rejection* (top). If, as in Fig. 67(a), the sample does not meet either of these lines, there is *further testing*.

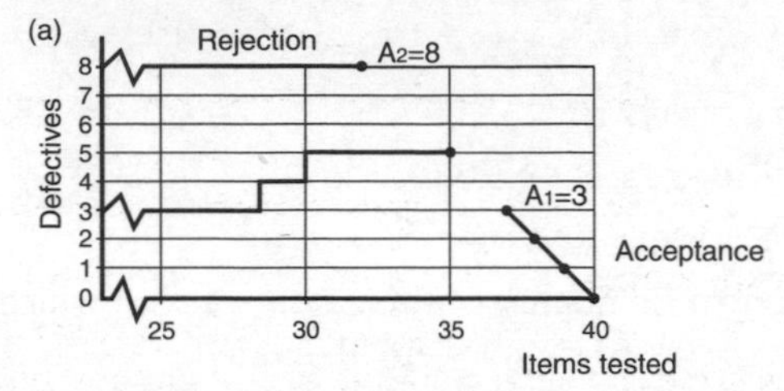

Fig. 67(a). **Lattice diagram.** (a) After a first sample of 40 items. There are 35 Good and 5 Defective.

In the case of further testing, a new acceptance number A_3 set (usually at the level of A_2 from the first sample, or higher) and a second sample is then tested.

Second sample: Size = 20, $A_3 = 7$

The batch will be accepted if the total outcome of sampling is one of:

(53 G, 7 D), (54 G, 6 D), (55 G, 5 D), (56 G, 4 D)

The results for the second are plotted on the same lattice diagram, Fig. 67(b). In this example there are 54 Good and 6 Defective items, so the batch is accepted.

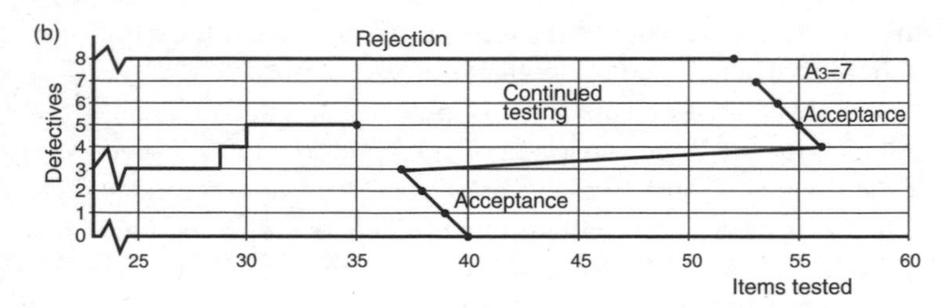

Fig. 67(b). **Lattice diagram.** (b) After a second sample batch of a further 20 items. There are 54 Good and 6 Defective items. The batch is accepted.

In a sequential sample, the lower line goes up in a series of steps, and sampling is continued until the sampling line for the batch enters either the acceptance or the rejection region. This is illustrated in Fig. 68.

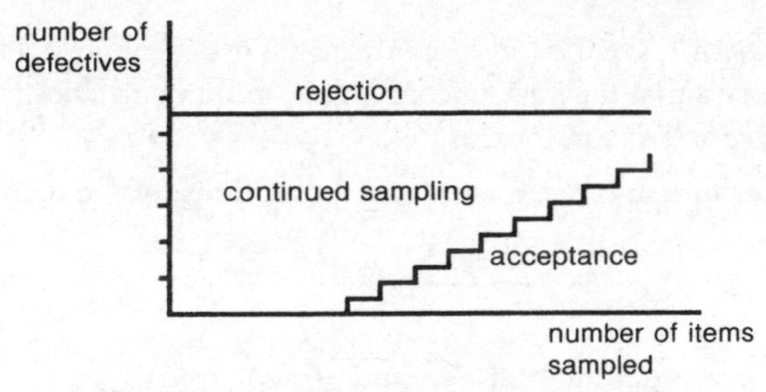

Fig. 68. **Lattice diagram sequential sample.**

law of large numbers the law which (in its simplest form) states that, as the size of samples is increased, their mean tends to a limit which is the same as the mean of the parent population from which the samples are drawn. This can be stated formally formally as

$$\text{For every } \varepsilon > 0,\ \mathrm{P}\left\{\left|\sum_{i=1}^{n}\frac{X_i}{n} - \mu\right| > \varepsilon\right\} \to 0 \text{ as } n \to \infty$$

where n is the sample size and μ is the population mean.

learning curve a graph of the results of an experiment involving learning, performance or skill, over a period of time or a number of tests.

least squares, method of the method of minimising the sum of the squares of the residuals, as a method of fitting models to data.

The method of least squares is illustrated here by the derivation of the formula for the y on x regression line. The sample data are a set of points $(x_1, y_1), (x_2, y_2), \dots, (x_n, y_n)$ and these are used to estimate the unknown population parameters α and β in the model

$$y_i = \alpha + \beta x_i + \varepsilon_i .$$

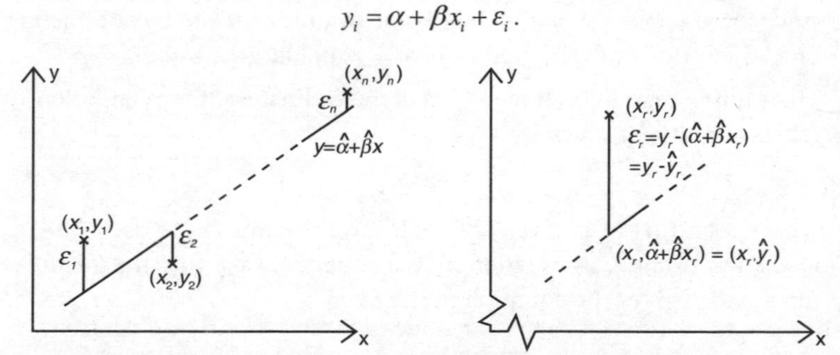

Fig. 69(a) and (b). **Least squares, method of.** Graph (a) illustrates 3 of the n data points, the regression line $y = \hat{\alpha} + \hat{\beta}x$ and the residuals ε_1, ε_2 and ε_n. Graph (b) shows the general point (x_r, g_r), the equivalent point on the regression line and the residual.

The values of ε_i are the random errors, the true (unknown) residuals. It is assumed that they are uncorrelated random variables with mean zero and constant variance.

The values of α and β are estimated by minimising the quantity

$$S=\sum_{i=1}^{n}\varepsilon_i^2=\sum_{i=1}^{n}\left[y_i-(\alpha+\beta x_i)\right]^2$$

This is done by setting both $\frac{\partial S}{\partial \alpha}$ and $\frac{\partial S}{\partial \beta}$ equal to 0.

This gives a pair of simultaneous equation in α and β but the values obtained by solving the equations for them are only estimates, based on the particular sample. So they are written $\hat{\alpha}$ and $\hat{\beta}$. They are given by:

$$\hat{\alpha}=\bar{y}-\hat{\beta}\bar{x} \text{ and } \hat{\beta}=\frac{\sum_{i=1}^{n}x_iy_i-n(\bar{x})(\bar{y})}{\sum_{i=1}^{n}x_i^2-n\bar{x}^2}.$$

It is usual to write $\hat{\beta}$ in the form $\hat{\beta}=\frac{S_{xy}}{S_{xx}}$ where

$$S_{xy}=\sum_{i=1}^{n}(x_i-\bar{x})(y_i-\bar{y})=\sum_{i=1}^{n}x_iy_i-n(\bar{x})(\bar{y}) \text{ and } S_{xx}=\sum_{i=1}^{n}(x_i-\bar{x})^2=\sum_{i=1}^{n}x_i^2-n\bar{x}^2.$$

It is, however, good practice to use the following alternative forms for S_{xx} and S_{xy}:

$$S_{xy}=\sum_{i=1}^{n}x_iy_i-\frac{1}{n}\sum_{i=1}^{n}x_i\sum_{i=1}^{n}y_i \text{ and } S_{xx}=\sum_{i=1}^{n}x_i^2-\frac{1}{n}\left(\sum_{i=1}^{n}x_i\right)^2$$

since these are less prone to the loss of accuracy caused by subtracting one large number from another nearly equal large number.

The sample regression line, which is the estimate of the population regression line, is given by:

$$y=\hat{\alpha}+\hat{\beta}x$$

where $\hat{\alpha}$ and $\hat{\beta}$ are as given above. It is quite common to use a and b instead of $\hat{\alpha}$ and $\hat{\beta}$ as notation in this equation. Inserting the formulae for $\hat{\alpha}$ and $\hat{\beta}$ gives the equation in the form

$$(y-\bar{y})=\frac{S_{xy}}{S_{xx}}(x-\bar{x}) \cdot$$

It is often easier to use it in this form.

likelihood the probability function (discrete case) or probability density function (continuous case) of a random variable X interpreted as a function of the parameter(s) given the value x of X, instead of as a function of x given the value(s) of the parameter(s).

Example: A shop has 6 £20 notes in its till. These may be of a new design or an old design. One note is chosen at random and inspected to see whether it is new design or old design.

The random variable, X, is the number of new design notes chosen when one note is selected. It can thus have values 0 and 1.

The number of new design notes in the till may be thought of as a parameter of the distribution of X. It may take the values 0, 1, 2, 3, 4, 5 or 6. For each value of the parameter the distribution of X is different. So when, for example, the parameter has value 2 (i.e. there are 2 new notes among the 6 in the till), $P(X = 0) = \frac{4}{6}$ and $P(X = 1) = \frac{2}{6}$. The full probability distribution for X in terms of x for the different parameter values is thus.

		Parameter value						
$P(X = x)$		0	1	2	3	4	5	6
x	0	1	$\frac{5}{6}$	$\frac{4}{6}$	$\frac{3}{6}$	$\frac{2}{6}$	$\frac{1}{6}$	0
	1	0	$\frac{1}{6}$	$\frac{2}{6}$	$\frac{3}{6}$	$\frac{4}{6}$	$\frac{5}{6}$	1

Now think of a case when the one note chosen is in fact a new design note. What is the likelihood distribution for the number of new notes among the 6 in the till?

Finding the likelihood distribution involves looking at the situation from a different perspective. Given that the one note chosen is of new design ($x = 1$), what are the probabilities that the number of new notes (the parameter value) is 0, 1, 2, 3, 4, 5 and 6.

For $x = 1$, the total of the probabilities for the different possible parameter values is

$$0 + \tfrac{1}{6} + \tfrac{2}{6} + \tfrac{3}{6} + \tfrac{4}{6} + \tfrac{5}{6} + 1 = \tfrac{21}{6}.$$

The likelihood distribution is then worked out as follows. It gives the probability of the particular parameter values, given the value of x (in this case 1).

Parameter value	Likelihood ($x = 1$)
0	$0 \div \frac{21}{6} = 0$
1	$\frac{1}{6} \div \frac{21}{6} = \frac{1}{21}$
2	$\frac{2}{6} \div \frac{21}{6} = \frac{2}{21}$
3	$\frac{3}{6} \div \frac{21}{6} = \frac{3}{21}$
4	$\frac{4}{6} \div \frac{21}{6} = \frac{4}{21}$
5	$\frac{5}{6} \div \frac{21}{6} = \frac{5}{21}$
6	$1 \div \frac{21}{6} = \frac{6}{21}$

If the note is not of new design, and so $x = 0$, the likelihood distribution would be as follows

Parameter value	Likelihood ($x = 0$)
0	$0 \div \frac{21}{6} = 0$
1	$\frac{1}{6} \div \frac{21}{6} = \frac{1}{21}$
2	$\frac{2}{6} \div \frac{21}{6} = \frac{2}{21}$
3	$\frac{3}{6} \div \frac{21}{6} = \frac{3}{21}$
4	$\frac{4}{6} \div \frac{21}{6} = \frac{4}{21}$
5	$\frac{5}{6} \div \frac{21}{6} = \frac{5}{21}$
6	$1 \div \frac{21}{6} = \frac{6}{21}$

In this example the parameter was the number of new notes in the till. It is more usual for the parameter to be one of those commonly associated with statistical distributions (mean, standard deviation, binomial probability, etc.).

Likelihood is often applied to the joint probability (density) function for a random sample.

See MAXIMUM LIKELIHOOD ESTIMATION.

linear estimator an ESTIMATOR which is a linear function of the sample values. For example, a linear estimator for the population mean from a random sample $X_1, X_2, \ldots X_n$ is given by:

$$a_1X_1 + a_2X_2 + \ldots + a_nX_n$$

where $a_1, a_2, \ldots, a_n$ are all constants and $a_1 + a_2 + a_n = 1$

In the case when $a_1 = a_2 = \ldots = a_n = \frac{1}{n}$, this becomes

$$\frac{X_1 + X_2 + \ldots + X_n}{n}$$

which is the sample mean, $\bar{X}$.

linear interpolation see INTERPOLATION.

linear model a model in which the observations are represented by linear combinations of parameters, plus experimental error. Such models include regression and analysis of variance models. Sometimes the expression *general linear model* is used.

linear relationship a relationship between two variables which gives rise to a straight line graph. In mathematics, the term implies an exact relationship whereas in statistics data are subject to random variation and so usually do not fit any relationship exactly. In this entry, for the sake of simplicity, the term linear relationship is described in the mathematical sense without random variation.

The relationship between the variables x and y is linear if it can be written in the form $y = mx + c$, where m is gradient and c is the intercept (Fig. 70).

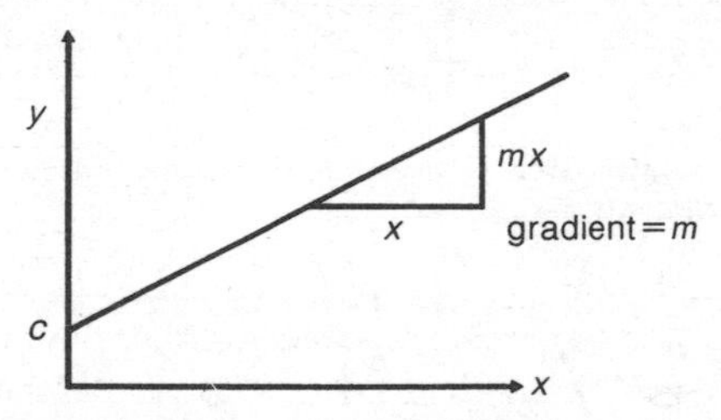

Fig. 70. **Linear relationship.**

Example: A company estimates the cost of printing n books to be £C where C = 2n + 5000. This is a linear relationship. £5000 is the cost of setting up the presses etc., and £2 the cost of printing each book thereafter.

It is sometimes possible to convert a non-linear relationship into a linear one, by algebraic manipulation and a change of variables. Examples are illustrated in Figs. 71, 72 and 73.

There are two major advantages in graphing a relationship in linear form. One is that it allows the equation to be found from the intercept and gradient of the graph. The other is that points which do not fit the relationship are easily spotted.

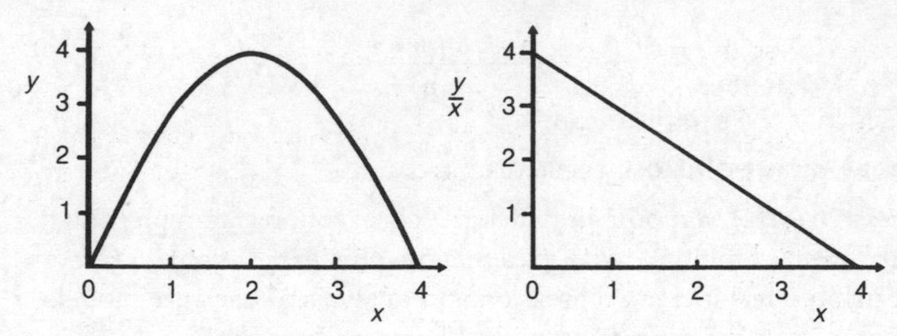

Fig. 71. **Linear relationship.** Conversion of non-linear relationship; $y = 4x - x^2$ becomes $\frac{y}{x} = 4 - x$.

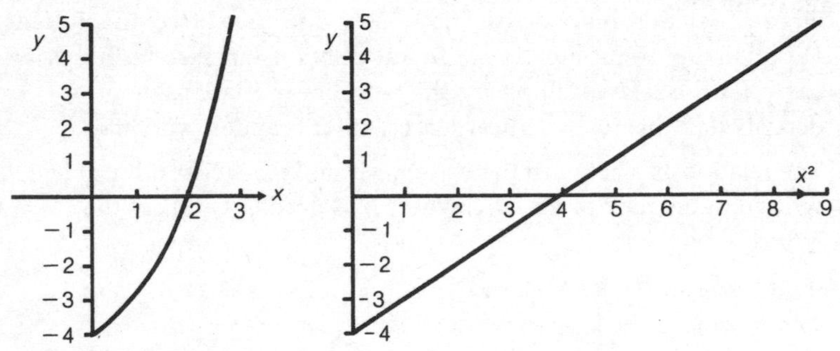

Fig. 72. **Linear relationship.** Conversion of non-linear relationship; $y = -4 + x^2$ is plotted as y against x^2.

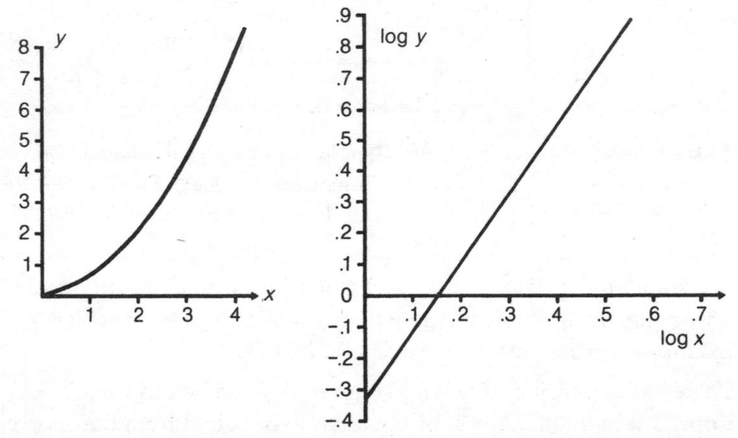

Fig. 73. **Linear relationship.** Conversion of non-linear relationship; $y = \frac{1}{2}x^2$ becomes $\log y = 2 \log x + \log \frac{1}{2}$.

line of best fit the straight line which fits a set of data most accurately when plotted on a graph.

A line of best fit may be drawn by eye, for example leaving the same number of points on each side of it. A more accurate line is provided by the regression line, but this does involve some assumptions about the variables involved. Usually the independent variable (see DEPENDENT VARIABLE) is plotted on the x-axis, being known accurately; the line of best fit is then the y on x regression line. (If the independent variable is plotted on the y-axis, the equivalent line of best fit is the x on y regression line.)

link function a function of the expected value of each variable in a generalised linear model which is equal to some linear combination of the parameters of the model. See GENERALISED LINEAR MODEL.

logarithm the power to which a fixed number, the base, must be raised to obtain a given number. Abbreviated to log. Thus the logarithm to the base a of the number b is defined as c in the equation $a^c = b$, and is written as:

$$c = \log_a b$$

Two bases are particularly important for logarithms, e and 10. Logarithms to the base e are called *natural logarithms*, and often denoted by ln ($\log_e$ is also correct). Natural logarithms arise because:

$$\int_1^X \tfrac{1}{x}\,dx = \ln X$$

Logarithms to the base 10 used to be widely used for quick calculations before electronic calculators were invented. They are still used in a number of circumstances, including log-log graphs (although natural logarithms would do equally well), and in the definitions of a number of quantities, e.g. pH value for acidity, the Richter scale for earthquakes, noise level in decibels.

For logarithms to any base a, the following relationships are true for $x > 0$, $y > 0$, $a > 1$.

$$\log_a(xy) = \log_a x + \log_a y$$

$$\log_a\left(\frac{x}{y}\right) = \log_a x - \log_a y$$

$$\log_a(x^n) = n \log_a x$$

$$\log_a(1) = 0$$

$$\log_a(a) = 1$$

All logarithmic graphs have the same shape, illustrated in Fig. 74 for base a ($a > 1$).

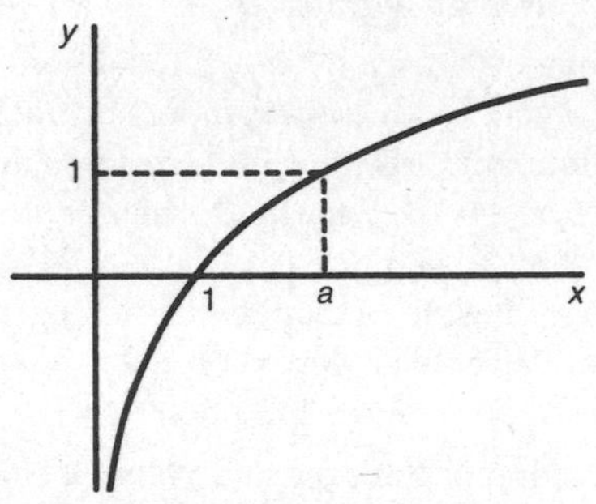

Fig. 74. **Logarithm.** Graph of $y = \log_a x$.

It is not possible to have the logarithm of a negative number.

As $x \to 0$, $\log_a x \to -\infty$.

As $x \to \infty$, $\log_a x \to \infty$.

The inverse of a logarithm is called its *antilogarithm* (antilog). If $y = \log_a x$, then $x = \text{antilog}_a\, y$. Thus $\text{antilog}_a\, y = a^y$.

logarithmic graph see GRAPH, LOGARITHM.

logistic model a model for the growth of a variable. In one form the equation has the form:

$$x = \frac{a}{1 + be^{-ct}}$$

The values of a, b and c are assigned for the particular variable under consideration; a represents the ultimate magnitude of the variable, c determines its rate of growth and and b (as well as a) is related to its initial value of $\frac{a}{1+b}$. Its graph is shown in Fig. 75.

The logistic curve is the solution of the differential equation

$$\frac{dz}{dt} = cz(1-z) \text{ where } z = \frac{x}{a};$$

the initial value of z is $\frac{1}{1+b}$.

This is one of many forms of the logistic model. Another form which is often used is given by $f(x) = \frac{e^{\alpha+\beta x}}{1+e^{\alpha+\beta x}}$.

In this $f(x) \to 1$ as $x \to \infty$, making it a useful model for probabilities or proportions.

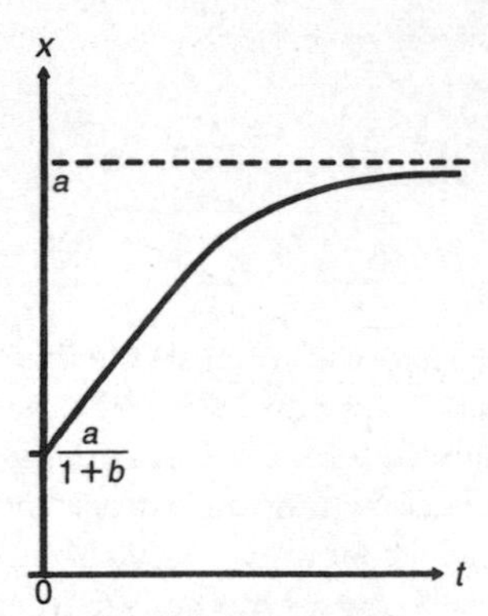

Fig. 75. **Logistic curve.**

logit a function, or transformation, which is sometimes applied to probabilities.

$$\text{logit}(p) = \log_e\left(\frac{p}{1-p}\right)$$

It transforms the scale for p from 0 to 1, to $-\infty$ to $+\infty$.

If the probability of an event occurring is p, the expression $\frac{p}{1-p}$ gives the odds on its occurring. For this reason logit is sometimes called the log odds function.

The logit function is used in logit (or logistic) regression modelling, in cases where it is more appropriate to express a probability model in the form logit(p) = ... rather than p = ...

Such a model could have the simple linear form $\text{logit}(p) = \alpha+\beta x$. This is a rearrangement of the logistic model $p = \frac{e^{\alpha+\beta x}}{1+e^{\alpha+\beta x}}$.

loglinear model a model for data in a 2-way contingency table in which the logarithm of the expected frequency in the cell (i, j) is expressed as an additive (linear function) combination of a row effect for row i, a column effect for column j and an interaction between them. This form of modelling extends to 3-way and higher-way contingency tables.

lot tolerance percentage defective, LTPD (in quality control) the maximum percentage of defectives accepted in a sampling scheme. If the percentage of defectives is no more then the LTPD, the batch from which the sample was drawn is accepted.

Malthus' theory (of historical interest) the theory that the human population will expand exponentially while the means of subsistence will increase more slowly, with the consequence that ultimately its size will be kept in check by famine, disease and war. It was put forward by the English economist T.R. Malthus (1766-1834). The widespread use of birth control methods since then has made much smaller rates of population growth possible; consequently the theory is now mainly of historical interest.

Mann Whitney *U* test a non-parametric test of the null hypothesis that two samples are drawn from populations with the same distributions, against the alternative hypothesis that they are drawn from different populations. The test is based on the location of the samples.

The Mann Whitney test is equivalent to the Wilcoxon Rank Sum 2-Sample Test, which may therefore be used as an alternative to it.

Example: two comparable groups of trainee pilots were taught the same course by different instructors. At the end of the course they took the same test, gaining these marks:

Lt.Leroy's group (L)	65	42	39	58	61	72		
Capt.French's group (F)	63	41	25	56	47	49	28	45

Is there evidence at the 5% significance level that one teacher is better than the other?

H_0: There is no difference in the marks according to teachers.
H_1: There is a difference according to teacher.

5% significance level
2-tail test

The value of the test statistic U is calculated as follows.

Take each F value and give it a score of how many of the L values it exceeds. Thus the first F value, 63, exceeds the L values, 42, 39, 58 and 61 and so has a score of 4. The test statistic, U, is the total of the scores for all these scores and so in this case is

$$4 + 1 + 0 + 2 + 2 + 2 + 0 + 2 = 13.$$

If the groups L and F are interchanged, a different value of U is obtained, in this case 35. The two values are often denoted U_1 and U_2 and $U_1 + U_2 = n_1n_2$, where n_1 and n_2 are the sizes of the two groups. Thus, in this case, 35+13 = 6×8 = 48. The test statistic, U, is the smaller of U_1 and U_2. In this case $U = 13$.

The null hypothesis is rejected if the test statistic U is less than the critical value.

The critical value for U is then found in the tables for the Mann Whitney test (Table 11). For a 2-tail test, for $n_1 = 6$, $n_2 = 8$, at 5% significance level, the critical values of U is 8.

Since 13 > 8, there are no grounds for rejecting the null hypothesis. There is no evidence, at the 5% level, to claim either instructor was better than the other.

The procedure given above for calculating the test statistic can be tedious and error prone for large samples. An alternative method is to rank the combined data. In the example above the combined ranks are as follows.

L	2	10	12	5	4	1		
F	3	11	14	6	8	7	13	9

The two values of U are then calculated using the formulae

$$n_1n_2 + \frac{n_1(n_1+1)}{2} - R_1 \text{ and } n_1n_2 + \frac{n_2(n_2+1)}{2} - R_2.$$

where R_1 and R_2 are the sums of the ranks of the two groups. The test statistic is the smaller of these two values.

If the value of n_1 or n_2 (or both) is greater than 25, it is too large to allow the tables to be used. In that case, the test can be carried out using a Normal approximation for U, with

$$Mean = \tfrac{1}{2}n_1n_2, \quad Standard\ deviation = \sqrt{\frac{n_1n_2(n_1+n_2+1)}{12}}.$$

A continuity correction is required when this approximation is used.

When ranks are tied the standard deviation is given by:

$$Standard\ deviation = \sqrt{\frac{n_1n_2}{N(N-1)}\left(\frac{N^3-N}{12} - \sum_i T_i\right)}$$

where $N = n_1 + n_2$ and $T_i = \dfrac{t_i^3 - t_i}{12}$ where t_i is the number of observations tied at rank i.

marginal distribution see BIVARIATE DISTRIBUTION.

Markov chain a sequence of events, the probability of each of which is dependent on the event immediately preceding it, but independent of earlier events. (Named after A. Markov, 1856-1922, Russian mathematician). Every Markov chain has a transisition matrix. See TRANSITION MATRIX.

Marshall Edgeworth Bowley Index see INDEX NUMBER.

matched groups groups which are matched for similarity before an experiment is carried out on one of them, the other being used as the control group (see CONTROL EXPERIMENT).

matched sample **1.** a sample in which the same attribute, or variable, is measured twice, under different circumstances.

2. two samples in which the members are clearly paired; see PAIRED SAMPLE.

maximum likelihood estimation a technique for estimating parameters in which the values that maximise the likelihood are chosen. Sometimes they can be found in a simple way using calculus, but at other times more complicated mathematics is required.

These are the "most likely" values of the parameters, given the data that have actually been observed. The method is of key importance in statistical inference. Maximum likelihood estimators possess many desirable properties, making them good (but not perfect) estimators in many ways.

mean **1.** A measure of an average value. There are several types of mean, used in appropriate circumstances but, unless stated otherwise, the term "mean" is usually taken to be the arithmetic mean.

For the numbers $x_1, x_2, ...x_n$:

$$\textit{Arithmetic mean} = \frac{x_1 + x_2 + ...x_n}{n}$$

$$\textit{Geometric mean} = \sqrt[n]{x_1 x_2 ... x_n}$$

$$\textit{Harmonic mean} = \frac{1}{\frac{1}{n}\left(\frac{1}{x_1} + \frac{1}{x_2} + ... + \frac{1}{x_n}\right)}$$

$$\textit{Weighted mean} = \frac{w_1 x_1 + w_2 x_2 + ...w_n x_n}{w_1 + w_2 + ... + w_n}$$

where $w_1, w_2, ...w_n$ are the weightings of the numbers $x_1, x_2, ...x_n$.

2. See EXPECTATION.

mean absolute deviation the mean of the absolute values of the differences between the values of a variable and the mean of its distribution. Thus, mean absolute deviation of $x_1, x_2, ...x_n$ is given by:

$$\sum_{i=1}^{n} \frac{|x_i - \bar{x}|}{n}.$$

See DISPERSION. Compare STANDARD DEVIATION, INTERQUARTILE RANGE.

mean square a quantity calculated in analysis of variance by dividing a sum of squares by its degrees of freedom. See ANALYSIS OF VARIANCE.

measure of central tendency a typical or central value of a variable, often referred to as an average. Commonly used measures of central tendency are mean (or average), mode, median, and midrange.

measures of dispersion see DISPERSION.

median the middle value in a distribution. The total frequency or probability below the median is equal to that above the median. For a set of n numbers, the median is that ranked $\frac{1}{2}(n+1)$. Thus the median of the values 13, 12, 10, 9, 15, 12, 8 is found after ranking the numbers as follows:

Number	*Rank*	
8	1	
9	2	
10	3	
12	4	←Median = 12
12	5	
13	6	
15	7	

(Notice that the ranking used to find the median is slightly different from that used at other times. When a tie occurs, as with the two 12s in the above example, they are ranked one above the other rather than equal.)

If there is an even number, the median is the arithmetic mean of those on either side of the middle. Thus the median of 1, 2, 3, 4, 5, 6 is 3.5.

median polish a method of fitting an additive model in a situation where the response variable is a function of two categorical variables. It is an Exploratory Data Analysis technique which, like a graph, makes it easy to look at the data and see what they have to say.

Example: a company wish to compare the yields obtained from a crop when they use one of five different fertilisers (A, B, C, D, E) and one of three different patterns of application (P, Q, R). In the applications, the total amount used is constant, but the timings are varied. Each possible combination of fertiliser and pattern is used once on plots of land which have been checked to be very similar. The results, in suitable units, are as follows:

				Fertiliser		
		A	**B**	**C**	**D**	**E**
	P	31	23	45	25	27
Pattern	**Q**	21	14	36	16	19
	R	26	21	36	31	32

The average effect of each of the fertilisers is taken as the median of the values in the columns. For fertiliser A it is 26. These medians are written in an extra row at the bottom of their columns. The original entries in the table are reduced by subtracting the column medians, as shown in the table below.

				Fertiliser		
		A	**B**	**C**	**D**	**E**
	P	5	2	9	0	0
Pattern	**Q**	-5	-7	0	-9	-8
	R	0	0	0	6	5
Column medians		***26***	***21***	***36***	***25***	***27***

This has removed the "fertiliser effects" enabling easier comparison of the patterns. For example, it is now clear that pattern Q is "below average", or at best "average", no matter which fertiliser is used.

The process is now repeated, starting with the new table, but using the rows. Thus the median of each row is identified, including the row of column medians, and these are written as an extra final column. Their values are subtracted from the entries in their rows.

		Fertiliser					*Row*
		A	**B**	**C**	**D**	**E**	*medians*
	P	3	0	7	-2	-2	**2**
Pattern	**Q**	2	0	7	-2	-1	**-7**
	R	0	0	0	6	5	***0***
Column medians		***0***	***-5***	***10***	***-1***	***1***	**26**

It is now possible to fit an additive model to the original data. Each value in the original table may be reconstructed as the sum of the overall average (26), the column median, the row median and the residual (the value remaining in the cell).

For example, the value in the bottom right hand corner, corresponding to fertiliser E and pattern R, was originally 32. This is now broken down as:

32 = 26 (overall effect) + 1 (fertiliser effect) + 0 (pattern effect) + 5 (residual)

This analysis provides a way of comparing the effects of the fertilisers with each other, so C (10) seems to be the best and B (-5) seems to be the worst.

It also gives a way of comparing the application patterns with each other, so P (2) seems to be the best and Q (-7) seems to be the worst.

If the differences between the treatments (the fertilisers and the patterns in this case) are large compared to the residuals (the values remaining in the final table), the conclusions are fairly reliable; otherwise they should be treated very cautiously. In the example, the range of the residuals is 9 (from -2 to 7) so the company should be wary of making much of any differences between the treatments which are less than this.

If one variable is of more interest than another, the procedure is to start by subtracting the medians of the value of less interest. Thus if, in the example, comparing the fertilisers is of greater interest, the company would start by subtracting the row medians.

The working so far is, in fact, one cycle/iteration of the median polish method. Working on the rows has affected the columns so that, for example, the median of the C column is now 7 instead of zero. The whole process may usefully be repeated, several times if necessary. Fewer alterations are usually necessary on each iteration. So, for the

C column 7 is added to the column median (giving 17) and 7 is subtracted from each of the entries in the column. Doing this for all of the columns gives:

		Fertiliser					*Row*
		A	**B**	**C**	**D**	**E**	*medians*
	P	1	0	0	0	-1	2
Pattern	**Q**	0	0	0	0	0	-7
	R	-2	0	-7	8	6	*0*
Column medians		*2*	*-5*	*17*	*-3*	*0*	*26*

The column of row medians would be included in this process, but in this case its median is already zero. In fact, all of the row medians are still zero, so the end of the process has been reached.

This analysis of the data has most of the residuals equal to zero, but the range is 15. The large residuals are all from pattern R.

meta analysis, the process of combining results from several different surveys or experiments.

If the surveys or experiments are basically identical then there is no problem in pooling the results. For example, if three interviewers each interview a random sample of people from the same area about the political party for which they intend voting in a forthcoming election, then it is reasonable just to pool their results.

If, on the other hand, the surveys are from different areas where the proportions voting for the different parties are usually very different, then it might not be appropriate just to pool the results. In more complicated cases such as this, a model is needed to reflect the expected differences between the sets of results, and this is used in combining them to give an overall picture.

method of least squares see LEAST SQUARES, METHOD OF.

method of moments a technique for estimating parameters.

The moments of a distribution are functions of its parameters. These functions can be inverted, to give the parameters as functions of the moments. The parameters are then estimated by the same functions of the sample moments.

This method is often simple and quick to apply, but it can be difficult to establish what, if any, desirable properties the resulting estimators possess.

method of semi-averages see SEMI-AVERAGES, METHOD OF.

midrange the arithmetic mean of the smallest and largest values in a data set. Thus, if $x_1, x_2, \ldots x_n$ are a set of numbers arranged in order of magnitude, then the midrange is

$$\frac{x_1 + x_n}{2}$$

It is half way between the largest and the smallest of the numbers.

Example: the midrange for the values 9, 9, 7, 7, 6, 5, 3, 3, 3, 2, is:

$$\frac{9+2}{2} = 5.5$$

The midrange of a sample is sometimes used as an estimator for the population mean. If the parent distribution is uniform, midrange is a more efficient estimator than the sample mean or the sample median, and is unbiased. See BIASED ESTIMATOR.

mixed model (in analysis of variance situations) an analysis of variance model in which some terms have the Model 2 interpretation and others do not. See MODEL 2.

mode the value of a variable which occurs most frequently.

Example: a biologist counts the eggs in each of 50 blackbirds' nests. The results were as follows:

Number of eggs	2	3	4	5	6
Frequency	1	10	26	11	2

The mode of the number of eggs is 4, which occurred 26 times; the next largest, 5, had a frequency of only 11.

If there are two modes, the distribution is said to be BIMODAL, three modes *trimodal*, and so on. The term, bimodal, is also applied to distributions with two distinct peaks even if they are not exactly equal in frequency.

For grouped data, the modal class is the class with the highest value of:

$$\frac{\text{Frequency}}{\text{Width of class interval}}$$

If the classes all have the same width of class interval the modal class is that with the most observations, the highest frequency. If however, the intervals are of different widths, this need not be the case. If a histogram is drawn, the modal class will, however, always be that with the highest rectangle.

Example: the numbers of hours, h, of sunshine per week over one year in a particular place were recorded, and illustrated on a histogram, as follows.

	$0 \leq h < 20$	$20 \leq h < 40$	$40 \leq h < 50$	$50 \leq h < 55$	$55 \leq h < 60$	$60 \leq h < 100$
Frequency	5	7	14	12	10	4

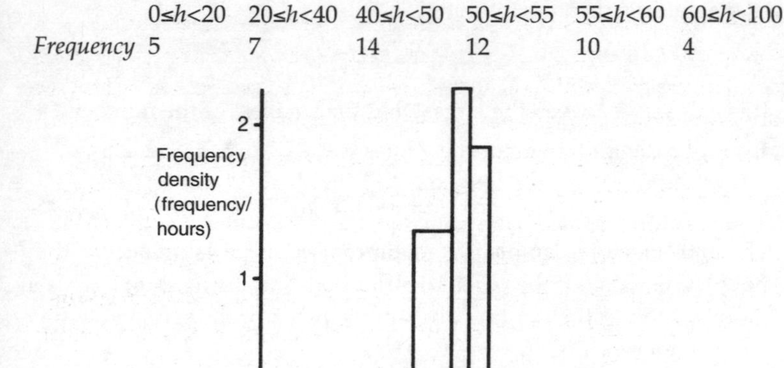

Fig. 76. **Modal class.** Frequency distribution of hours of sunshine per week for a period of one year.

The modal class is not 40-50, even though this has the highest frequency. The class 50-55 has a higher frequency density as shown on the histogram (Fig. 76), and so this is the modal class.

model see MODELLING.

Model 1 see MODEL 2.

Model 2 (in analysis of variance situations) an analysis of variance model in which the levels of the factors are interpreted as a random sample from a population of possible levels.

Example: in a one-way analysis of variance, the performances of operatives in carrying out a production line task are being compared. The sensible interpretation is that the performance of individual operatives for their own sake is of no interest, but rather that they are representative of a population of potential operatives who could be employed to carry out the particular task. So information is wanted about the variation in that population. This is a *Model 2* situation.

In contrast, in a comparison of the effects of some fertilisers on the yield of a crop, the particular fertilisers would probably be the only subject of interest, with no suggestion of a wider background population of fertilisers. This is the more customary interpretation of analysis of variance models, and is sometimes known as *Model 1*.

Other variations of modelling interpretations also sometimes arise.

modelling *v.* the process of representing a real-life situation in a mathematical form or as a set of procedures (an algorithm). The result is called a *model* and this may be qualified by a description of the type of model, for example *mathematical, statistical* or *computer*.

A typical real-life situation is complicated and so most modelling involves simplification in order to make it tractable. In a modelling cycle, the model is tested against reality and the simplification is progressively lessened until the model is judged good enough.

Statistical modelling often involves selecting a standard distribution and finding values for its parameters so that it provides a good match for observed data. Known properties of the model distribution (e.g. its mean, variance, skewness, median) are then used to make predictions about the real situation.

modulus see ABSOLUTE VALUE.

moment the rth moment of a distribution of the random variable X about the point a is defined by $\mathrm{E}\left[(X-a)^r\right]$. See EXPECTATION.

If the random variable X is discrete, the rth moment about a is given by

$$\sum_i (x_i - a)^r \mathrm{p}(x_i).$$

If the random variable X is continuous, it is given by

$$\int_{\text{all } x} (x-a)^r \mathrm{f}(x)\mathrm{d}x$$

Although a can have any value, it is usual to take a either as zero (giving *moments about zero* or *moments about the origin*, notation μ'_r) or as the mean of the distribution (giving *moments about the mean*, notation μ_r).

It can be seen that, about the mean;

the 0th moment, $\mu_0 = 1$

the 1st moment, $\mu_1 = 0$ (since μ is the mean)

the 2nd moment, $\mu_2 =$ the variance.

The 3rd and 4th moments about the means, μ_3 and μ_4, are used in measures of skewness and kurtosis respectively. See SKEW, KURTOSIS.

Example: Find the 1st, 2nd, 3rd and 4th moments of the data set 1, 2, 4, 5, 8, 10 about the mean.

The mean $\bar{x} = \frac{30}{6} = 5$

	x	$(x-\bar{x})$	$(x-\bar{x})^2$	$(x-\bar{x})^3$	$(x-\bar{x})^4$
	1	−4	16	−64	256
	2	−3	9	−27	81
	4	−1	1	−1	1
	5	0	0	0	0
	8	3	9	27	81
	10	5	25	125	625
Σ	30	0	60	−60	1044

So the moments are as follows 1st moment $\mu_1 = \frac{0}{6} = 0$

2nd moment $\mu_2 = \frac{60}{6} = 10$

3rd moment $\mu_3 = \frac{-60}{6} = -10$

4th moment $\mu_4 = \frac{1044}{6} = 174$

For moments about the origin,

the 0th moment, $\mu'_0 = 1$

the 1st moment, $\mu'_1 = Mean, \mu$

the 2nd moment, $\mu'_2 = Variance + \mu^2$

For sample data, $x_1, x_2, ..., x_n$, the rth moment about a is given by $\frac{1}{n}\sum_{i=1}^{n}(x_i - a)^r$. The commonly occurring cases are $a = 0$ and $a = \bar{x}$, often denoted by m'_r and m_r respectively.

moment generating function the moment generating function of the distribution of the random variable X is given by:

$E(e^{Xt}) = \sum_i e^{x_i t} p(x_i)$ (for a discrete variable)

$= \int_{\text{all } x} e^{xt} f(x) dx$ (for a continuous variable).

where $p(x_i)$ is the probability distribution and $f(x)$ the probability density function. Symbol: $M_X(t)$.

When these expressions are expanded using

$$e^{Xt} = 1 + Xt + \frac{X^2t^2}{2!} + \frac{X^3t^3}{3!} + ...$$

the results can be separated into

$$M_X(t) = \mu'_0 + \mu'_1 t + \mu'_2 \frac{t^2}{2!} + ... + \mu'_n \frac{t^n}{n!} + ...$$

where μ'_n is the nth moment of the distribution about the origin. Thus the moment generating function, written as a power series expansion, gives all the moments (about the origin) of the random variable. (Note: these moments can also be obtained by repeatedly differentiating $M_X(t)$ and setting $t = 0$; often this is easier.)

For independent random variables *A* and *B*, with moment generating functions $M_A(t)$ and $M_B(t)$, the moment generating function for the variable $(A + B)$ is given by:

$$M_{A+B}(t) = M_A(t) \times M_B(t).$$

The probability distributions (or probability density functions) of two random variables are the same if, and only if, they have the same moment generating function. Thus the moment generating function completely specifies a distribution. Not all distributions however possess moment generating functions and, for this reason, the *characteristic function*, $E(e^{iXt})$ is defined, where i is the complex number $\sqrt{-1}$. This exists uniquely for every probability distribution.

Monte Carlo method a method of finding the probability distribution, or some other characteristic, of the possible outcomes of a process or experiment by simulation. (Named after the casino at Monte Carlo, where systems for winning at games of chance such as roulette, etc., are often tried.)

There are processes which are too complicated to allow a theoretical analysis. In such cases, the probabilities of the various outcomes may be estimated by carrying out repeated simulations.

Example: A new supermarket is being designed. One of the questions to be considered is how many cash points will be needed if long queues are not to build up at the busiest times. Rather than attempt a theoretical analysis of the situation, the designers carry out computer simulations with different numbers of cash points. For each one they are able to build up the probability distributions for different lengths of queues, 0, 1, 2, ... at the cash points.

moving average a form of average which has been adjusted to allow for seasonal components or cyclic components. Some variables, for example the number of unemployed people, have considerable seasonal or cyclic components. These may make it difficult to determine the underlying trend. These components can be eliminated by taking a suitable moving average.

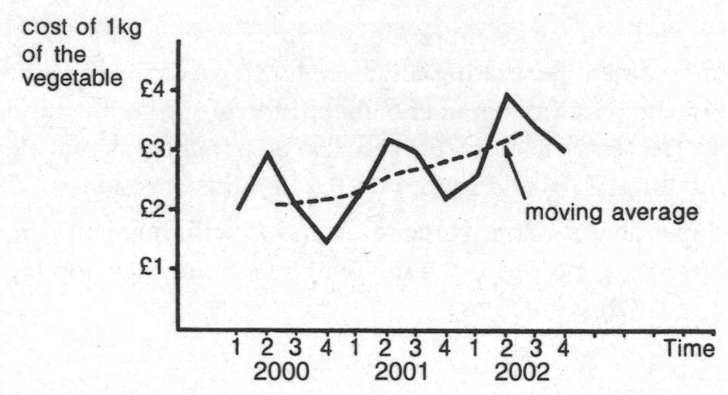

Fig. 77. **Moving average.** The cost of a tropical vegetable plotted for each quarter year, and the moving average.

Example: the cost of a kilogram of a tropical vegetable in London (averaged over each quarter) over the years 2000-2002 was as follows:

	2000	*2001*	*2002*
1st quarter	£2.00	£2.20	£2.60
2nd quarter	£3.00	£3.20	£4.00
3rd quarter	£2.00	£3.00	£3.40
4th quarter	£1.40	£2.20	£3.00

The seasonal component, which is very noticeable on the graph (Fig. 77), is eliminated when the *four-point moving average* is taken. The calculation for this is set out below, and the resulting average plotted on the graph (Fig. 77).

Start	'00,1	'00,2	'00.3	'00,4	'01,1	'01,2	'01,3	'01,4	'02,1
End	'00,4	'01,1	'01,2	'01,3	'01,4	'02,1	'02,2	'02,3	'02,4
Mean time	'00,2½	'00,3½	'01,½	'01,1½	'01,2½	'01,3½	'02,½	'02,1½	'02,2½
	2.00	3.00	2.00	1.40	2.20	3.20	3.00	2.20	2.60
Price	3.00	2.00	1.40	2.20	3.20	3.00	2.20	2.60	4.00
	2.00	1.40	2.20	3.20	3.00	2.20	2.60	4.00	3.40
	1.40	2.20	3.20	3.00	2.20	2.60	4.00	3.40	3.00
Total	8.40	8.60	8.80	9.80	10.60	11.00	11.80	12.20	13.00
Mean	2.10	2.15	2.20	2.45	2.65	2.75	2.95	3.05	3.25

This shows a clear upward trend in the cost of the vegetable.

A problem with the four term moving average in the example is that its values are centred on imaginary points, 2½, 3½ and so on, where there are no real data values to compare them with. This can be overcome by taking the averages of successive values, a two-term moving average of the terms of the four-term moving average. This gives a weighted moving average over five quarters, $\frac{1}{8}$ (1,2,2,2,1), since:

$$\frac{1}{2}\left[\frac{1}{4}(a_1 + a_2 + a_3 + a_4) + \frac{1}{4}(a_2 + a_3 + a_4 + a_5)\right]$$
$$= \frac{1}{8}(a_1 + 2a_2 + 2a_3 + 2a_4 + a_5)$$

where a_1, a_2, a_3, a_4 and a_5 are five successive quarterly values. This value is centred at point 3.

The number of parts in a moving average is clearly related to the time scale involved, and should be such that each average is over one complete cycle of variation. This is often, but by no means always, one year; studies of long-term changes in sun-spot frequency, for example, would require moving averages to be taken over an 11-year cycle.

multi-level modelling a sophisticated form of modelling that endeavours to account for variation at different levels in a structure. It is quite widely used in educational and medical research, in particular. For example, in comparing performances of school pupils, it might be sensible to model variation between local authority (or other geographical) areas, between schools within local authorities and between classes within schools.

multiple regression a technique for fitting a regression model for a random variable which is dependent on several other variables.

Such a model for a random variable V which is dependent on several variables x_1, x_2, … is

$$V = \alpha_1 x_1 + \alpha_2 x_2 + \ldots + \varepsilon$$

where α_1, α_2, ... are constants and ε represents the residual variation. The values of α_1, α_2, ... are estimated on the basis of sample data, usually by the method of least squares.

multiplication rule the rule which states that, for two events A and B, the probability p of their both occurring is given by:

$$\mathrm{P}(A \cap B) = \mathrm{P}(A|B).\mathrm{P}(B)$$
$$\text{or} \quad \mathrm{P}(A \cap B) = \mathrm{P}(B|A).\mathrm{P}(A)$$

If the events A and B are independent, this can be simplified to

$$\mathrm{P}(A \cap B) = \mathrm{P}(A).\mathrm{P}(B)$$

Example: a man goes to the races and backs a horse, entirely at random, on each of the 2.00 and 2.45 races. The first race has five runners, the second seven. What is the probability that he picks both winners?

The first event, 'He picks a winner in the 2.00 race', has probability $\frac{1}{5}$; the second event, 'He picks a winner in the 2.45 race', has probability $\frac{1}{7}$.

The two events are independent. The probability that he picks both winners is

$$\tfrac{1}{5} \times \tfrac{1}{7} = \tfrac{1}{35}$$

multiplicative model a model in which the individual terms are multiplied together, rather than added (as is more usual).

multivariate analysis techniques for analysing multivariate distributions. Such techniques may be divided into two categories: ordination and regression. The main aim of ordination techniques is to reduce the dimension of the data (for example by classifying some variables together). The aim of regression techniques is to model the variables so that the value of one can be predicted from the values of the others.

See REGRESSION MODEL.

multivariate distribution a distribution involving a number of distinct, but not necessarily independent, variables. If two variables are involved it is called bivariate. The number of variables is called the *dimension*. Compare BIVARIATE DISTRIBUTION.

Example: The joint distribution of height, base diameter and age for the trees in a beech forest is multivariate.

multivariate regression the extension of multiple regression (which involves a random variable Y on several non-random variables, x_i) to the situation where the x_i variables are random too.

See MULTIPLE REGRESSION.

mutually exclusive see EXCLUSIVE.

natural logarithm see LOGARITHM.

negative binomial distribution also called Pascal's distribution, the probability distribution given by:

$$p(x) = \binom{x-1}{n-1} p^n q^{x-n}$$

for $x \geq n$ where $0 < p < 1$ and $q = 1 - p$. This is the probability distribution for the number of independent Bernoulli trials needed to achieve n successes. Each trial has probability p of success, q of failure. The mean and variance of this distribution are given by:

$$Mean = \frac{n}{p} \qquad Variance = \frac{nq}{p^2}$$

In the case when $n = 1$, the negative binomial distribution is the geometric distribution.

nested model also called hierarchical model or hierarchal model, an analysis of variance model having a structure in which some terms are naturally located (or nested) within other terms.

Example: a manufacturer of diesel engines has two factories (*A* and *B*) each of which produces the same four types of engine (I, II, III, IV). An experiment is conducted to compare their reliabilities. The model will need a set of terms to represent the performances of the four engine types. However, there might be a consistent overall difference between factory *A* and factory *B* for all the engine types; the model needs terms for the possible factory-to-factory difference, and the possible differences between engine types are nested within the factory differences.

nominal scale see CATEGORICAL SCALE.

non-inferior see SUPERIOR.

non-linear interpolation see INTERPOLATION.

non-parametric test of significance or **distribution-free test of significance** a test of significance which makes no assumptions concerning the underlying distribution and the values of any associated parameters. All tests involving the ranks of data are non-parametric. Examples are Kendall's rank correlation and Spearman's rank correlation (see CORRELATION COEFFICIENT), Kruskal-Wallis one-way analysis of variance, Friedman's two-way analysis of variance by rank, Mann Whitney *U* test, the sign test, and the Wilcoxon tests. Compare PARAMETRIC TEST OF SIGNIFICANCE.

norm an average level of achievement, performance or behaviour.

Normal curve, Gaussian curve or **error curve** the graph of the probability density function of the Normal distribution. It is a bell-shaped curve (Fig. 78), symmetrical about its mode (or mean, or median) Symbol: $\varphi(x)$.

Normal distribution a continuous distribution of a random variable with its mean, median and mode equal. Notation: $N(\mu,\sigma^2)$ where μ is the mean and σ^2 the variance.

The probability density function $\varphi(x)$ of the normal distribution, with mean μ and standard deviation σ, is given by:

$$\varphi(x) = \left(\frac{1}{\sqrt{2\pi}\sigma}\right) e^{\left\{-\frac{(x-\mu)^2}{2\sigma^2}\right\}}$$

The graph of the probability density function of the Normal distribution is the continuous bell-shaped curve illustrated in Fig 78. This curve is also called the Normal curve or *error curve* and is symmetrical about the mean.

A Normal distribution is often standardised by the transformation

$$Z = \frac{X-\mu}{\sigma}$$

so that it has mean 0 and standard deviation 1. The standardised distribution is denoted by N(0, 1), and has probability density function given by:

$$\varphi(z) = \frac{1}{\sqrt{2\pi}} e^{-\frac{1}{2}z^2}.$$

The points on the curve are given in Table 1.

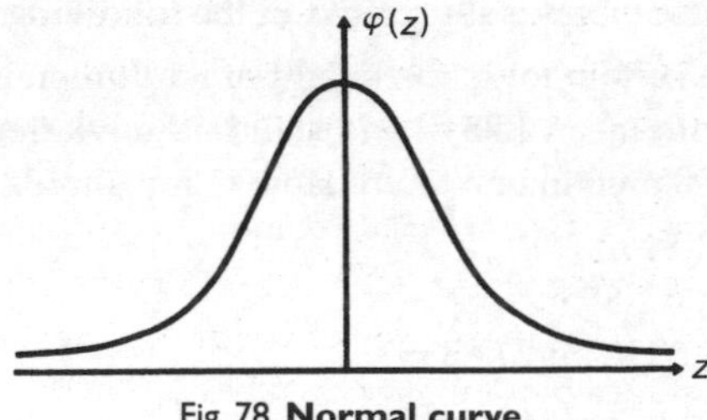

Fig. 78. **Normal curve.**

The Normal distribution is very important in statistics for two reasons:

(a) It is the distribution of many naturally-occurring variables, such as the heights of adult men in a region, the masses of carrots in a field, etc.

(b) The Central Limit Theorem. This states that the distribution of the means of samples drawn from any parent population, whether Normal or not, is itself Normal or approximately so. As the sample size increases, this distribution approaches the Normal distribution with increasing accuracy.

The area under the Normal curve represents the proportion of the population in question. It can be found from tables but using them requires some care since they may be given in various ways, as illustrated in Fig. 79. Table 2 is of type (a), and the area in question in this case is denoted by $\Phi(z)$.

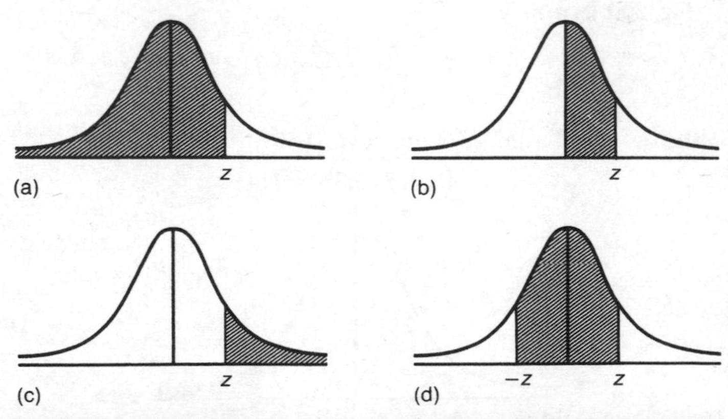

Fig. 79. **Normal distribution.** Graphs showing areas sometimes given in tables.

The use of these tables is illustrated in the following example.

Example: in a certain town the height of adult men is Normally distributed with mean 1.75 m and standard deviation 0.05 m. There are 5,000 men in one town. How many should be:

(a) less than 1.80 m;

(b) greater than 1.85 m;

(c) between 1.80 m and 1.85 m?

(a) The area required is found as

$$\Phi\left(\frac{1.80-1.75}{0.05}\right)=\Phi(1)=0.8413$$

It is represented by the shaded area in Fig. 80(a). Since the probability of a man being less than 1.80 m tall is 0.8413, the predicted number of men in this category is:

$$0.8413\times 5{,}000 = 4{,}206$$

(a) 1.75 1.80

Fig. 80(a). **Normal distribution.** (a) Area for z < 1.80.

(b) For men taller than 1.85 m, the area required is that shaded in Fig. 80(b), namely:

$$1-\Phi\left(\frac{1.85-1.75}{0.05}\right)=0.0228$$

So the predicted number of men over 1.85 m tall is:

$$0.0228\times 5{,}000 = 114$$

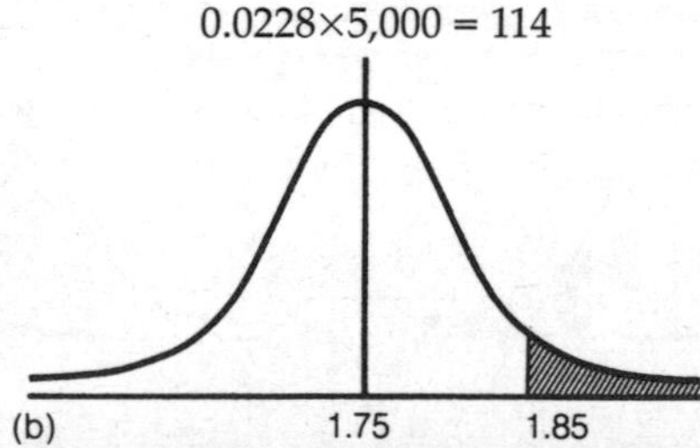

Fig. 80(b). **Normal distribution.** (b) Area for z > 1.85.

(c) The number of men between 1.80 m and 1.85 m in height (see Fig. 71(c)) is predicted to be:

$$5{,}000-(4{,}206 + 114) = 680$$

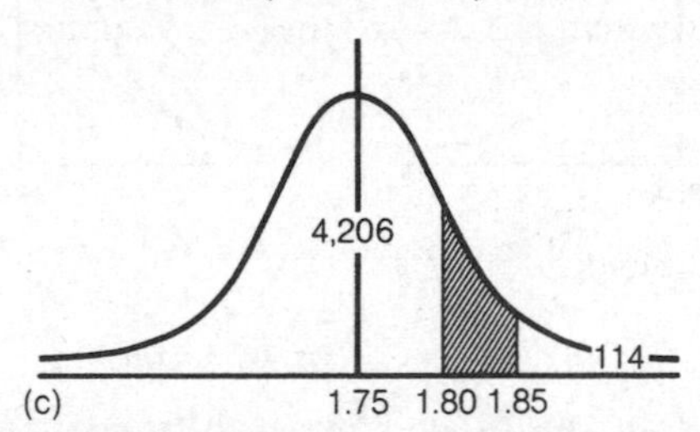

Fig. 80(c). **Normal distribution.** (c) Area for 1.80 < z < 1.85.

A commonly used hypothesis test involves the mean of a Normal distribution. This is illustrated in the following example.

Example: a company manufacturing light-bulbs claim that their bulbs are designed to have a mean life of 800 hours and standard deviation 40 hours. A customer finds that 50 bulbs that he bought had mean life 790 hours, and claims that they were substandard because their mean life was too small. The manufacturer replies that there is nothing wrong with his bulbs, but that the customer picked a bad sample. Assuming the standard deviation of the life of the bulbs is indeed 40 hours, test the customer's claim at the 5% significance level.

$H_0 \quad \mu = 800.$
$H_1 \quad \mu < 800.$

5% significance level
1-tail test

If the null hypothesis is true, the distribution of samples of size 50 has:

$$Mean = \mu = 800,\ Standard\ deviation = \frac{\sigma}{\sqrt{n}} = \frac{40}{\sqrt{50}} = 5.657$$

The area shaded in Fig. 81 represents the probability that a random sample has mean ≤790. It is given by:

$$\Phi\left(\frac{790-800}{5.657}\right) = \Phi(-1.77)$$

By the symmetry of the Normal curve, this area (Fig. 81(a)) is the same as that to the right of 810 (Fig. 81(b)). This is $1 - \Phi(-1.77) = 0.0384$.

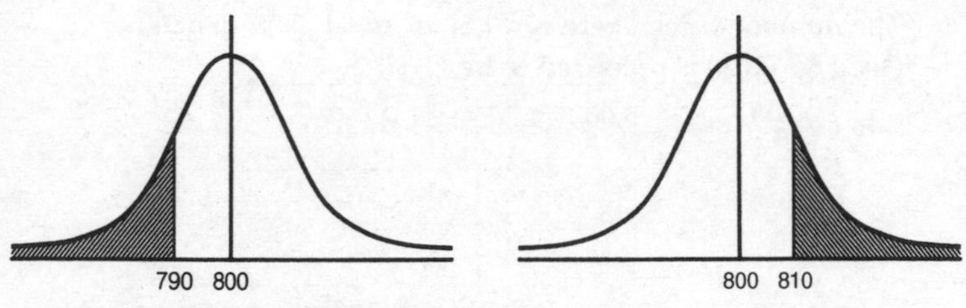

Fig. 81(a) and (b). **Normal distribution.** (a) Area for mean < 790. (b) Area for mean > 810.

Thus, if the null hypothesis is true, the probability of a sample having a mean as low as, or lower than, that recorded is 0.0384.

Since 0.0384 < 5%, the null hypothesis is rejected. There are reasonable grounds for being suspicious of the manufacturer's claim.

A useful way of checking if sample data may have been drawn from a Normal distribution is to use *Normal probability graph paper*.

Example: 100 specimens of the common shrew were collected and weighed. Their masses were found to be as follows:

Mass, m (g)	$5.0 \le m < 5.5$	$5.5 \le m < 6.0$	$6.0 \le m < 6.5$	$6.5 \le m < 7.0$
Frequency (shrews)	1	10	24	31
Mass, m (g)	$7.0 \le m < 7.5$	$7.5 \le m < 8.0$	$8.0 \le m < 8.5$	$8.5 \le m < 9.0$
Frequency (shrews)	18	10	4	2

The first step is to draw up a cumulative frequency table, and then convert the cumulative frequencies into proportions of the total.

Mass, m (g)	$m<5.0$	$m<5.5$	$m<6.0$	$m<6.5$	$m<7.0$
Cum. fr.	0	1	11	35	66
Proportion	0	0.01	0.11	0.35	0.66
Mass, m (g)	$m<7.5$	$m<8.0$	$m<8.5$	$m<9.0$	
Cum. fr.	84	94	98	100	
Proportion	0.84	0.94	0.98	1.00	

These points are then plotted on special Normal probability graph paper (excepting the two end points, 0 and 1, which cannot be plotted); see Fig. 82. If the result is a straight line, the distribution is Normal. (Notice, however, that real statistical data are always subject to random variability and so the points would not be expected to lie exactly on a straight line.) In this example, the distribution is not Normal, but not very far from it either.

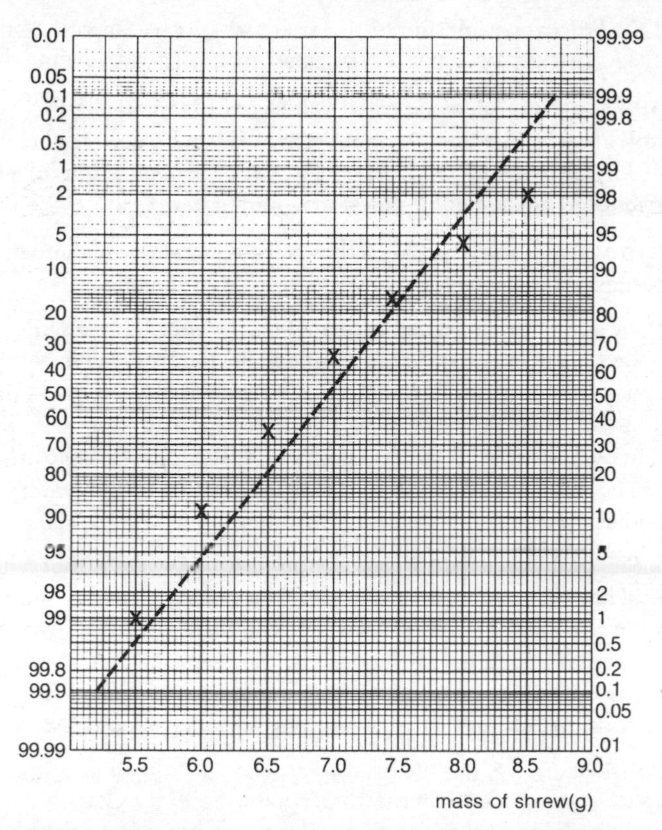

Fig. 82. **Normal distribution.** An example of data plotted on normal probability graph paper. The dashed line represents a Normal distribution, here, the data show positively skew, but with nearly Normal distribution.

The shape of the curve of a non-Normal distribution gives an indication of its skewness. The distribution of the shrews extends further to the right of the mean than to the left, and so has positive skewness. Numerical tests of how well data match the Normal distribution can be carried out using the Kolmogorov-Smirnov test or the chi-squared (χ^2) test.

The Normal distribution is often used as an approximation to other distributions. Examples are the binomial distribution when the number of trials is large and the probability neither very small nor close to 1, the Poisson distribution when the parameter is not small, and the Wilcoxon rank sum test . See KOLMOGOROV-SMIRNOV TESTS, BINOMIAL DISTRIBUTION, POISSON DISTRIBUTION, WILCOXON TESTS.

Normal scores a set of expected values of order statistics for samples from a Normal distribution, which can be used as the basis for various non-parametric tests. Order statistics from other distributions are sometimes used in a similar way.

null hypothesis the hypothesis that is tested against an alternative hypothesis in a statistical hypothesis test. Notation: H_0.

Statistical hypothesis testing is carried out by setting up a null hypothesis, H_0, and an *alternative hypothesis*, H_1. Data are then obtained from one or more samples. The probability that data, or more extreme results, are a chance result of the sampling, with the null hypothesis true, is then worked out. If this probability turns out to be smaller than the significance level of the test, the null hypothesis is rejected in favour of the alternative hypothesis.

A null hypothesis is usually based upon the assumption that nothing special has occurred, no change has taken place. When one sample is involved, the null hypothesis is often that the sample is drawn from a given parent population; for more than one sample, that the samples are drawn from the same parent population as each other. The alternative hypothesis is that something special has occurred, or a change has taken place.

There are two possible alternative hypotheses; that a change has taken place (leading to a two-tail test), and that a change has taken place in a particular direction (leading to a one-tail test); see ONE- AND TWO-TAIL TESTS.

Example: Mr and Mrs McTaggart have eight daughters and no sons. Can anyone say that they are medically special?

The null hypothesis is that there is nothing special about them. They are equally likely to have both boys or girls. It is just by chance that the children are all girls.

This is written

H_0: $p = \frac{1}{2}$.

where p is the probability that a baby is a girl.

The null hypothesis is tested against an alternative hypothesis. In this example there are, at least in theory, 3 possible alternative hypothesis.

$$p < \tfrac{1}{2}$$
$$p > \tfrac{1}{2}$$
$$p \neq \tfrac{1}{2}$$

The first of these does not match the context but either of the other two could be chosen as the alternative hypothesis. Either

H_1 $p > \tfrac{1}{2}$ (1-tail test).

or

H_1 $p \neq \tfrac{1}{2}$ (2-tail test).

To complete the specification of the of the hypothesis test, the significance level must be stated. Only then may the test itself be carried out.

The null hypotheses for a number of common tests are given below:

Test	H_o
Binomial probability test	The population probability of success is a particular value.
Normal test for mean	The population mean has a particular value.
t test for a mean	The population mean has a particular value.
Pearson's Product Moment Correlation Coefficient	$\rho = 0$.
χ^2 test	The data are drawn from a population with a given distribution.
χ^2 test for variance	The population variance has a particular value.
F test	The two populations have the same variance.

numerical data data where the items are recorded as numbers, for example the weights of the fish in a lake in kg, rather than in categories, for example the types of the fish. Numerical data may be continuous or discrete. See CATEGORICAL DATA, CONTINUOUS VARIABLE, DISCRETE VARIABLE.

O

odds the ratio of the probability of an event occurring to that of its not occurring.

Example: if the odds against Bolton Wanderers winning the FA cup are judged to be 1:20, this means:

P(Bolton win): P(Bolton do not win) = 1:20

and so

P(Bolton win) = $\frac{1}{21}$
P(Bolton do not win) = $\frac{20}{21}$.

Example: during an outbreak of a particular disease, it is observed that the probability of an individual contracting it is $\frac{1}{5}$. This means that the odds against someone contracting it are given by:

P(the person contracts the disease): P(the person does not contract the disease) = $\frac{1}{5} : \frac{4}{5}$ = 1:4. See LOGIT.

ogive a cumulative frequency graph. See CUMULATIVE FREQUENCY.

one- and two-tail tests tests of a null hypothesis against different alternative hypotheses.

Statistical hypothesis testing is carried out by setting up a null hypothesis and testing it against an alternative hypothesis.

This often involves testing the null hypothesis that a population parameter, like the mean or variance, has a particular value. For such tests, there are three types of alternative hypothesis:

(a) that the value of the parameter is different from that in the null hypothesis;

(b) that the value of the parameter is greater than that in the null hypothesis;

(c) that the value of the parameter is less than that in the null hypothesis.

Fig 83 illustrates possible critical regions for a test. They correspond to two tails of the distribution of the test statistic.

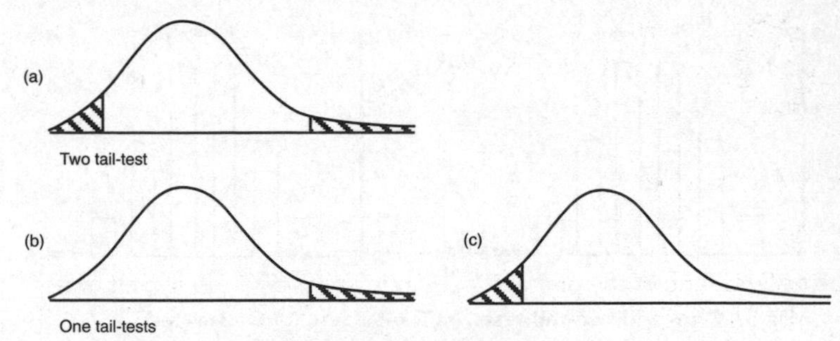

Fig. 83. **One- and two-tail tests.** Possible critical regions for a hypothesis test.

In case (a), the null hypothesis is rejected if the test statistic falls in either tail and so this is described as a two-tail test.

In case (b), the null hypothesis is rejected if the test statistic falls in the right hand tail (but is accepted if it falls in the left hand tail). So this is a one-tail test.

Similarly, case (c) is also a one-tail test but in that case the test statistic must fall in the left hand tail if the null hypothesis is to be rejected.

Example: a coin is tossed seven times, coming heads all seven, so that its fairness is called into doubt. Test, at the 1% significance level, (a) if it is biased; (b) if it is biased towards heads.

In both cases the null hypothesis is that the coin is unbiased. The probability of a head at any toss is $\frac{1}{2}$.

$H_0 \quad p = \frac{1}{2}$

For (a), the alternative hypothesis is that the probability of heads is not $\frac{1}{2}$, leading to a 2-tail test.

$H_1 \quad p \neq \frac{1}{2}$ (2-tail test)

For (b), the alternative hypothesis is that the probability of heads is greater than $\frac{1}{2}$, leading to a one-tail test.

$H_1 \quad p > \frac{1}{2}$ (1-tail test).

The two situations are illustrated in Fig. 84.

For (a), the probability of a result as extreme as 7 heads is the probability of 7 heads or 7 tails:

$$\left(\tfrac{1}{2}\right)^7 + \left(\tfrac{1}{2}\right)^7 = 0.0156$$

Since this is greater than 1%, there is no reason to reject the null hypothesis at this level. It cannot be claimed that the coin is biased at the 1% significance level.

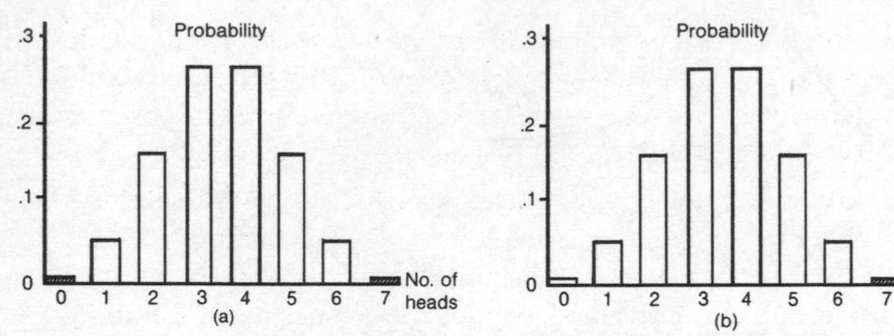

Fig. 84. **One- and two-tail tests.** (a) Two-tail test. (b) One-tail test.

For (b), the probability of 7 heads is:

$$\left(\tfrac{1}{2}\right)^7 = 0.0078$$

Since this is less than 1%, the null hypothesis is rejected. At this level the evidence supports the claim that the coin is biased towards heads.

This example shows that the outcome of a test can be affected by whether it is a 1- or 2-tail. The decision on whether to use a 1- or 2-tail test depends on the alternative being tested and this should be decided before collecting the data rather than when they are known.

It should be noted that both 1- and 2-tail tests are not available for all hypothesis tests. The description above applies to tests on the values of population parameters. Not all tests fit this description. Goodness-of-fit tests, for example, do not. In such cases it may be that the only possibility is a 1-tail test, as is usually the case with the χ^2 test.

one-way analysis of variance see ANAYLSIS OF VARIANCE, KRUSKAL-WALLIS ONE-WAY ANALYSIS OF VARIANCE.

open question see CLOSED QUESTION.

operating characteristic a function, often plotted as a graph, giving the probability of accepting the null hypothesis in a significance test. It is a function of the parameter under investigation.

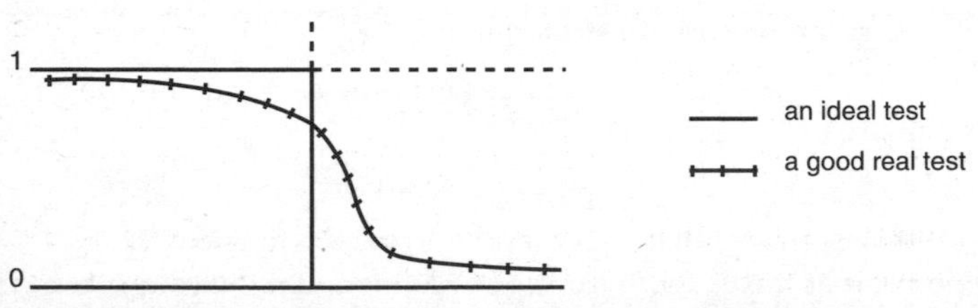

Fig. 85. **Operating characteristic.** An ideal test and a good real test.

The concept is illustrated in Fig. 85 for the test of $H_0 : \mu = \mu_0$ against $H_1 : \mu > \mu_0$. The continuous straight lines illustrate the operating characteristic of an ideal test (no such test exists); superimposed on it is a curve showing how a good real test might be expected to behave.

In quality control, the operating characteristic curve is the graph of the probability of accepting a batch from which a sample has been drawn for inspection, against the true proportion of defectives for this batch (Fig. 86). This curve is dependent on the particular sampling scheme being used.

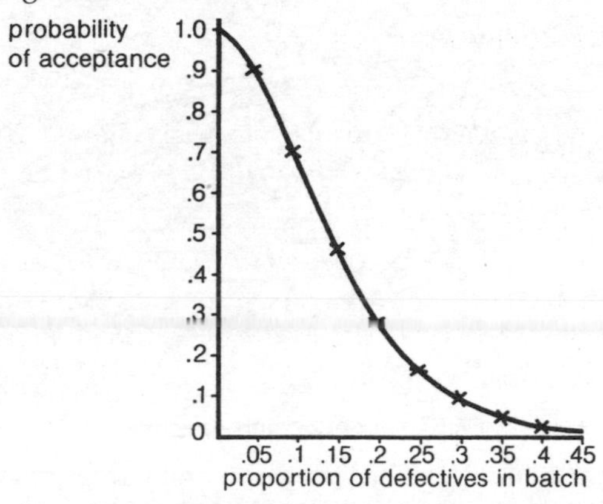

Fig. 85. **Operating characteristic curve.** The operating characteristic curve for a sampling scheme in which the sample size is 12, and the acceptance number is 1 (the batch is accepted if 0 or 1 are faulty).

operational research The analysis of problems in business, industry and other organisations using mathematical and statistical techniques.

opportunity sampling a method of sampling used when more representative methods are not possible.

The intention in drawing a sample from a population is that it should be representative of that population. Usually the best way of achieving this is to take a random sample. However, this requires a sampling frame. If a sampling frame is not available, and nothing is known about the structure of the population, then an opportunity sample may be taken. This is essentially any sample that may be conveniently obtained. It could, for example, involve interviewing a group of people who have assembled for some other purpose. The interviewer may try to obtain some kind of cross section of the population, or may try to eliminate personal bias by interviewing, say, every tenth person. Many samples are obtained this way, because it is not possible to obtain a sampling frame. However, the results should be treated with caution.

If some structure in the population is known, it is usual to try to take this into account in the sampling method, for example by using QUOTA SAMPLING.

Fig. 87. **An opportunity sample.**

or used to join alternatives. The word *or* is ambiguous in English and so needs to be used carefully, particularly in connection with probability calculations. For example, the condition, 'If he is over 2 m tall or over 100 kg in weight…' can mean:

(a) If he is over 2 m tall, or over 100 kg in weight, or both; in this case the original *or* is inclusive because the possibility of both events occurring is included;

(b) If he is either over 2 m tall, or over 100 kg in weight, but not both. The *or* is now exclusive, because the case when both events occur is excluded.

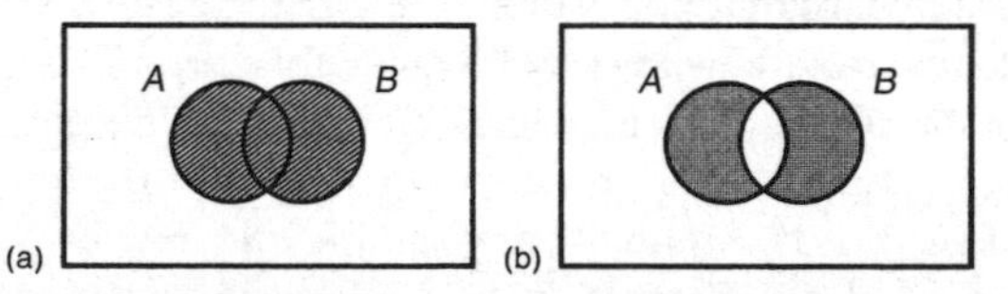

Fig. 88. **Or.** Venn diagram to illustrate the inclusive meaning of *or*, $A \cup B$.
Fig. 89. **Or.** (a) Venn diagram to illustrate the exclusive meaning of *or*, $(A \cup B) \cap (A \cap B)'$.

For events A and B, these two meanings can be illustrated on Venn diagrams, as in Figs. 88 and 89.

order statistics an ordering of data. Suppose that a random sample is denoted by $x_1, x_2, ..., x_n$. It is often convenient to arrange these in ascending order; they are then usually denoted by $x_{(1)}, x_{(2)}, ..., x_{(n)}$ and these are called the *order statistics* for the original sample.

Example: the sample 28, 24, 36, 22, 48 has $x_{(1)} = 22$, $x_{(2)} = 24$, $x_{(3)} = 28$, $x_{(4)} = 36$ and $x_{(5)} = 48$.

Order statistics are often important in the theory of non-parametric tests. See NORMAL SCORES.

ordinal number a number donating order, quality or degree in a group, such as first, second, third. Compare CARDINAL NUMBER.

ordinal scale a scale with classification and rank. However, equal differences in position on the scale do not necessarily represent equal differences in the variable being used for classification.

Example: a political party is selecting a candidate for an election and conducts a popularity poll among the electorate for the five contenders. The result is as follows:

Allotey	29%
Appiah	32%
Mensah	2%
Twumasi	30%
Zwennes	7%

They are then set in order:

1	Appiah
2	Twumasi
3	Allotey
4	Zwennes
5	Mensah

This 1-to-5 scale is an ordinal scale. Number 1 is more popular than number 2 and so on. However, the difference between numbers 1 and 3 (3%) is not the same as that between numbers 3 and 5 (27%).

ordinate see ABSCISSA, GRAPH.

outcome **1.** the result of an experiment or other situation involving uncertainty.

2. a sample point, an element in a sample space. See EVENT.

outlier an observation which is far removed from the others in the set. An outlier may be an observation from a different parent population, or it may be the result of experimental error, or it may be a genuine result. Since outliers can have a considerable influence on test statistics, they should be examined carefully before being accepted.

Two common criteria for deciding whether a point should be regarded as an outlier are that it is more than 2 standard deviations away from the mean, or that it is more than 1.5 × the interquartile range away from the nearer quartile.

The diagram (Fig. 90) represents a number of stars plotted according their emission of visible light and their surface temperatures on logarithmic scales. (Such a graph is called a *Hertzsprung-Russell Diagram.*)

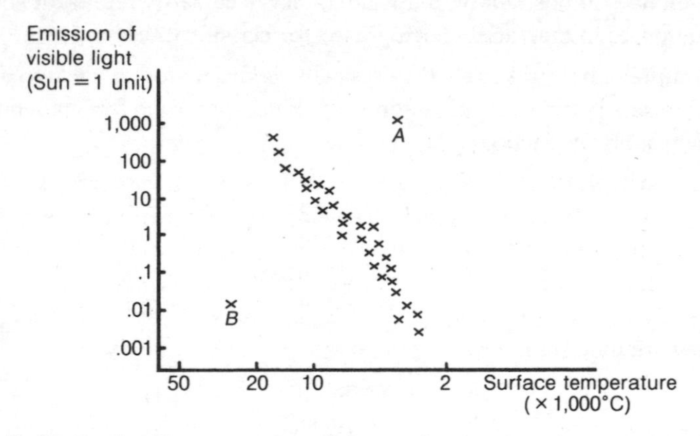

Fig. 90. **Outlier.** Hertzsprung-Russell diagram for a number of stars. *A* and *B* are outliers.

Clearly the stars *A* and *B* are outliers. The reason for this is that all the other stars are in the phase of their evolution when they are burning hydrogen into helium; they are main sequence stars. Star *A*, however, is in a later stage of its evolution, and is a red giant. Star *B* is in a still later stage and is a white dwarf. Thus, in this case, the two outliers belong to essentially different populations. See EXPLORATORY DATA ANALYSIS. Compare INLIER.

Paasche's Index see INDEX NUMBER.

paired sample two samples in which the same attribute, or variable, of each individual member of the sample is measured twice, under different circumstances. The term is also used in the case of two samples in which the members of each are clearly paired.

Examples include the times of a groups of athletes for 1500 m before and after a week of special training; the milk yields of the members of a herd of cows before and after being fed on a particular diet; intelligent quotient measurements on pairs of identical twins, and measurements of wear on left and right shoes.

Paired sample tests are carried out on the difference in the value of the variable for each matched pair.

paired sample test a test on the differences of the random variable measured in a paired sample. In a paired sample test on the means, the null hypothesis is that the mean value of the difference is a particular value; that value is often zero. Although two samples (usually before treatment and after it) are taken, the test is carried out on the single sample of the differences in the observations for each of the subjects.

In this entry the "before treatment" sample is denoted by $x_1, x_2, \dots, x_n$. The equivalent "after treatment" sample is denoted by $y_1, y_2, \dots, y_n$ and the differences by $d_1, d_2, \dots, d_n$. Notice that the values x_1, y_1, and d_1 all refer to the same individual, similarly for x_2, y_2, and d_2 and so on. The mean and estimated standard deviation of the population of differences are denoted by μ_d and s_d respectively.

The example which follows illustrates one type of paired sample test: the t test for the difference in means.

Example: the egg yields of a random sample of ten hens, labelled H_1 to H_{10}, are measured over a four-week period. The hens are then fed a new type of pellet and their yields are recorded over the next four weeks. The data are given below. Can it be claimed, at the 5% significance level, that the results show that the pellets alter hens' egg yield?

H_0: $\mu_d = 0$
H_1: $\mu_d \neq 0$

5% significance level
2-tail test

	Hen, H_i	H_1	H_2	H_3	H_4	H_5	H_6	H_7	H_8	H_9	H_{10}
YIELD	Before, x_i	26	24	15	23	14	18	17	28	18	19
(eggs)	After, y_i	28	24	18	21	19	21	17	27	24	26
	Difference, $d_i = y_i - x_i$	+2	0	+3	-2	+5	+3	0	-1	+6	+7

The mean of the observed differences is

$$\bar{d} = \frac{2+0+3-2+5+3+0-1+6+7}{10} = 2.3$$

Estimated population standard deviation

$$s_d = \sqrt{\left(\frac{(-0.3)^2+(-2.3)^2+(0.7)^2+(-4.3)^2+(2.7)^2+(0.7)^2+(-2.3)^2+(-3.3)^2+(3.7)^2+(4.7)^2}{10-1}\right)}$$

$$= 3.06$$

The estimated standard deviation of the means of samples of size 10 is

$$\frac{s_d}{\sqrt{n}} = \frac{3.06}{\sqrt{10}} = 0.967$$

The test statistic is given by

$$t = \frac{\bar{d} - \mu_d}{\frac{s_d}{\sqrt{n}}} = \frac{2.3-0}{0.967} = 2.38$$

Critical value: 2.26 (See Table 3 for 10-1 = 9 degrees of freedom for a 2-tail test).

Since 2.38 > 2.26 (the critical value), the null hypothesis is rejected at the 5% level, and it is concluded that the hens' egg yield is altered by the new pellets. However, since the values 2.38 and 2.26 are close it might be wise to conduct further tests to confirm this conclusion.

Notice that this paired sample t test with small samples assumes that the population of the differences is Normal.

The table below gives a number of common paired sample tests,

Test	H_o	Test Statistic
Normal test on the mean of a paired sample	The mean of the differences is μ_d	$\dfrac{\bar{d}-\mu_d}{\frac{\sigma}{\sqrt{n}}}$
t test on the mean of a paired sample	The mean of the differences is μ_d	$\dfrac{\bar{d}-\mu_d}{\frac{s_d}{\sqrt{n}}}$
Sign test	The median of the differences in the matched pairs is zero	The difference in the number of cases in which the difference between the pairs in the sample is positive and the number of cases in which it is negative
Wilcoxon matched-pairs Rank Sum Test	The median of the differences in the matched pairs is zero.	W = Sum of negative signed ranks of $\lvert$rank - median$\rvert$

Care must always be taken not to confuse a paired-sample test with a two-sample test. A two-sample test is used when the individual members of the two samples are not paired.

paradox a seemingly absurd or self-contradictory statement.

parameter **1.** a quantity used in defining the distribution of a population.

A Normal distribution is defined by two parameters, its mean and variance. So the Normal distribution with mean 20 and variance 7 is written N (20, 7).

A binomial distribution is also defined by two parameters, the number of trials n and the probability, p, of success in each one. It is written B(n, p).

A Poisson distribution is defined by just one parameter, the mean (or variance) λ and is written Poisson(λ).

Compare STATISTIC.

2. A quantity used in defining a model.

parametric test of significance a hypothesis test which assumes that the population has a particular type of distribution (e.g. Normal) and tests the value of one of the population parameters. The parameter tested depends on the distribution.

Some examples of parametric tests are: the Normal test and the t test for the population mean, the use of a correlation coefficient for the population correlation and the χ^2 test for the population variance.

Compare NON-PARAMETRIC TEST OF SIGNIFICANCE.

parent population the population from which a sample is drawn; usually this is just referred to as the *population*. See also SAMPLE SPACE.

There are times when an underlying parent population may be assumed, even though it does not strictly exist. The recovery times of 20 people who have contracted a new disease form a sample from the (non-existent) parent population which would be formed if many other people caught the disease.

Pareto charts *n* a means of prioritising actions in quality control when something is going wrong in a process, such as on a production line.

Example: a company makes printers. They are receiving a lot of complaints about a particular model of printer and many of them are being returned for repairs. The senior management could simply tell everyone that they should do better and take more care. However, it would be much more helpful if they were to focus on particular areas. A first move is to collect the data on the complaints and put them into classes such as: scratches on the case, paper jamming, poor print quality, etc. They are listed in order of frequency.

Fault	Number	Percentage	Cumulative percentage
Paper jams	97	49.7	49.7
Poor print quality	53	27.2	76.9
Print cartridge jam	25	12.8	89.7
Scratches	14	7.2	96.9
Power failure	3	1.5	98.5
Switch not working	3	1.5	100.0

Showing these in a bar chart, as in Fig. 91, highlights which areas are the source of most complaints. Putting the bars in decreasing order of size makes this even clearer.

A further step is to set a cut-off point for the areas on which to focus attention. The cumulative percentage of complaints is found, working from the largest problem area towards the smallest, and stopping when this exceeds 80%. (According to the 80-20 rule, this will usually involve work on about 20% of causes.) This is illustrated on the Pareto chart below. So, in this case, the management should focus attention on solving the paper jams, the poor print quality and the print cartridge jam.

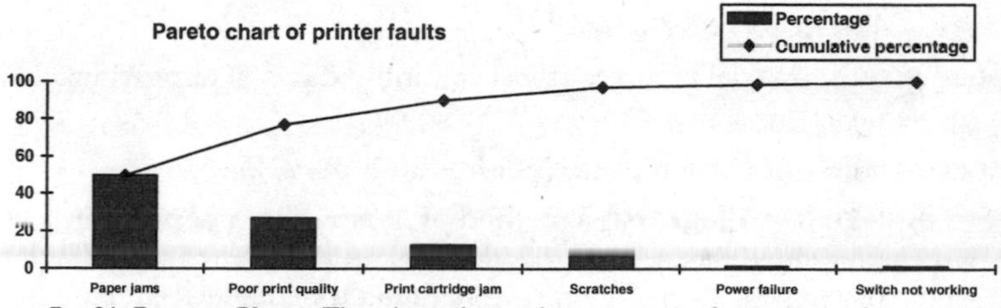

Fig. 91. **Pareto Chart.** This chart shows the percentage frequency and cumulative frequency of reported faults.

An alternative approach to using the Pareto chart is to use the cost of sorting out the problem instead of the frequencies. The first approach would probably be more important to the workers on the production line and the second to the management. In practice, a combination of the two would usually be used.

Pascal's distribution see NEGATIVE BINOMIAL DISTRIBUTION.

Pascal's triangle a triangular diagram consisting of rows of numbers (Fig. 92). It is named after Blaise Pascal (1623-62), French philosopher and mathematician. Each row is formed from the one above it by putting 1 at each end, and then placing new numbers at regular intervals, each calculated by adding the two numbers on either side of it on the row above.

1
1 1
1 2 1
1 3 3 1
1 4 6 4 1
1 5 10 10 5 1
1 6 15 20 15 6 1
• • • • • • • •

Fig. 92. **Pascal's triangle.**

Example:

. . . .
. . 5 10 . .
. . 15 . . 15 = 5 + 10

Pascal's triangle may be used for calculating binomial coefficients. This relationship is illustrated by comparing row four of Pascal's triangle:

1 4 6 4 1

and the binomial expansion for $(x + y)^4$:

$$(x + y)^4 = 1x^4 + 4x^3y + 6x^2y^2 + 4xy^3 + 1y^4$$

Pearson's measure of skewness see SKEW.

Pearson's product-moment correlation coefficient see CORRELATION COEFFICIENT.

peer group a social group composed of individuals of approximately the same age or status.

percentage proportion or rate per hundred parts. Symbol: %.

percentage bar diagram a method of illustrating a population which is divided into several discrete sets. In this sort of diagram, a single bar represents the whole population (100%) and is divided into parts in the correct proportions (Fig. 93). Percentage bar diagrams may be drawn horizontally or vertically.

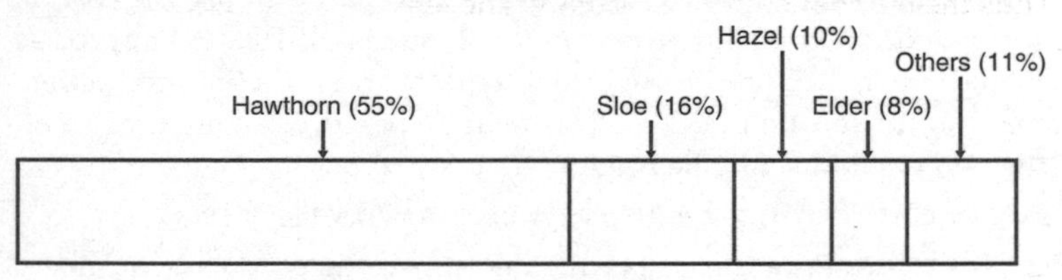

Fig. 93. **Percentage bar diagram.** The plants found in an old hedge.

percentage frequency distribution a frequency distribution expressed as a percentage of the whole.

Example: daily rainfall on one year at a weather station.

Rainfall, r (cm)	$0 \le r < 0.25$	$0.25 \le r < 0.5$	$0.5 \le r < 1.0$	$1.0 \le r < 2.0$
Frequency	157	66	89	53
% frequency (nearest 1%)	43	18	24	15

percentile one of 99 actual or notional values of a variable, which, with the two end values, divide its distribution into 100 groups with equal frequencies. See also QUANTILE.

permutation an ordered arrangement of the elements of a set, or of a subset of given size. The number of permutations of n elements, all, different is $n!$ Thus there are $3! = 6$ permutations of the letters A, B, C, namely:

$$ABC \quad ACB \quad BCA \quad BAC \quad CAB \quad CBA$$

The number of permutations of n elements selected r at a time is denoted by ${}^nP_r$ and given by

$${}^nP_r = \frac{n!}{(n-r)!}.$$

Thus there are $\frac{5!}{(5-2)!} = 20$ permutations of 2 letters chosen from the 5 letters A, B, C, D and E:

AB AC AD AE *BA BC BD BE* *CA CB CD CE*
DA DB DC DE *EA EB EC ED.*

The number of permutations of n objects, p of one type, q of another, r of another, etc. is given by

$$\frac{n!}{p!q!r!\ldots}$$

Thus the number of permutations of the letters P O S S E S S E S is

$$\frac{9!}{5!\ 2!\ 1!\ 1!} = 1512$$

The term permutation is often used incorrectly by those doing the football pools to describe selections where the correct term is combination.

pictogram a diagram in which a picture or icon is used to depict a given quantity of an item. The frequency of that item is indicated by the number of icons, all of which are identical (except that part icons may be used). See Fig. 94.

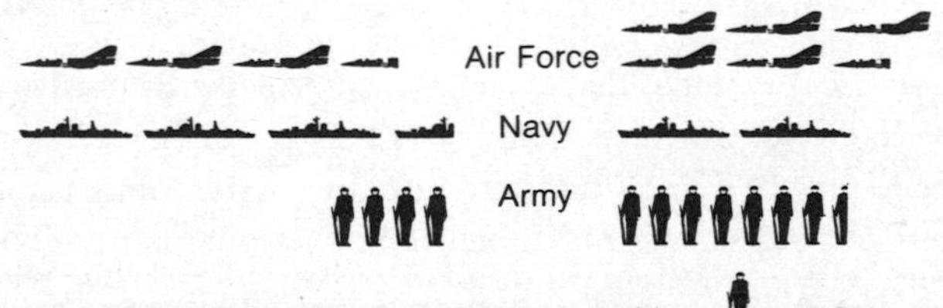

Fig. 94. **Pictogram.** The forces of two warring states.

A common mistake is to attempt to show frequency by the size of one of the icons: the problem with doing this is that it is unclear whether the ratio of the frequency to the number depicted by the basic icon is the scale factor for lengths, areas or volumes.

pie chart a circular diagram divided into sectors proportional to the frequencies of the items represented.

Example: a company spends £6 million on advertising one year, allocated as follows:

Television	£3 million
Sports sponsorship	£1 million
Newspaper adverts	£1.5 million
Posters	£0.5 million

This is illustrated by the pie chart in Fig. 95. The total of £6 million is represented by 360° and so £1 million is represented by a sector of 60°. The angles for the different sectors are thus:

Television	$3 \times 60^\circ = 180^\circ$
Sports sponsorship	$1 \times 60^\circ = 60^\circ$
Newspaper adverts	$1.5 \times 60^\circ = 90^\circ$
Posters	$0.5 \times 60^\circ = 30^\circ$.

Fig. 95. **Pie chart.**

Pie charts are sometimes displayed as discs; such a diagram is called a *slit chart*, Fig 96. However, many statisticians feel that it is bad practice to use slit charts since they introduce a third dimension, the thickness of the disc, but this dimension carries no information.

When two or more pie charts are drawn for comparison, their areas must be proportional to the sizes of the populations they are

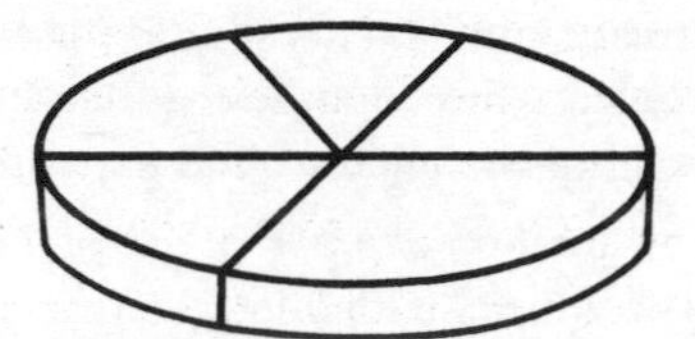

Fig. 96. **Pie chart.** A slit chart.

representing, and not their radii. The two pie charts in Fig. 97 represent the populations of a village in 1990 and 2010. Their radii are in the ratio 5:6 and so the ratio of their areas is $\pi \times 5^2 : \pi \times 6^2 = 25:36$. The population in 1990 was 1000 and so that in 2010 is given by:

$$\frac{36}{25} \times 1000 = 1440.$$

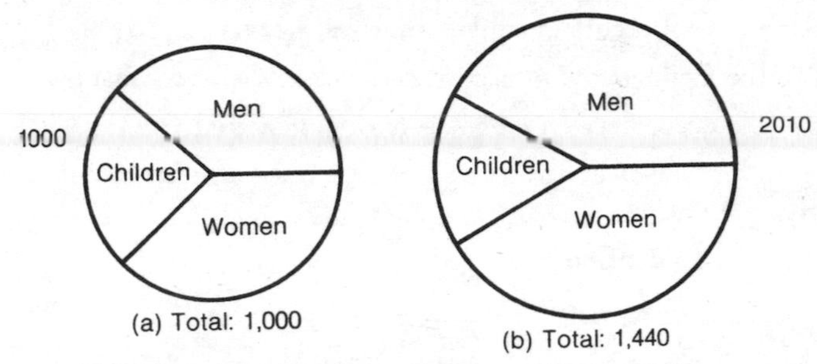

Fig.97. **Pie chart.** Population of a village in 1990 and 2010.

pilot survey a survey carried out, usually on a small scale, before a larger survey. It is carried out to check the appropriateness of the intended procedures, particularly with regard to sampling and design of questionnaires. If little is known in advance, the term *exploratory survey* is sometimes used.

placebo a treatment which is administered in experiments on humans, knowing that it should have no effect, such as a sugar pill. When a treatment is being tested on humans, there is often a psychological effect. A patient may improve simply because of what he or she believes is happening, rather than because of the effect of the medicine. To investigate the effect of "no treatment", some patients are given a placebo in the form of "medication" which is known to have no physical effect. This allows the extra non-psychological effect to be measured for those patients who are receiving the real treatment. Patients should not be aware of which of the treatments they are receiving, so the placebo should appear the same to the patients. This is called a blind experiment.

point estimation estimation of a parameter of a population as a single value. This contrasts with *interval estimation*, where the parameter is estimated to lie within an interval. Thus the sample 40, 52, 68, 70, 80 could be used to estimate the population mean as 62 (the sample mean). This would be a point estimate. The population mean could also be estimated as lying, say between 42.3 and 81.7, the 95% confidence interval; that would be an interval estimate.

point quadrat see QUADRAT.

Poisson distribution the probability distribution of a discrete random variable, X, representing the number of times (0, 1, 2, ...) that an event occurs in a given interval, when

(i) the average rate of occurrences is known, and constant,

(ii) the occurrences happen at random instants.

The interval is usually a period of time but may also be a distance, area, volume, etc.

The Poisson distribution is given by

$$P(X = r) = \frac{e^{-\lambda}\lambda^r}{r!}$$

for $r = 0, 1, 2, \ldots$.

The variance of the Poisson distribution is equal to its mean and this quantity is usually referred to as the Poisson parameter. This parameter is often denoted by the symbol λ, as given in the formula above, and the corresponding distribution is denoted by Poisson(λ).

Example: a football team scores 84 goals in 42 matches. The probability distribution of its goals per match may be modelled by the Poisson distribution with an estimated value of λ of $\frac{84}{42} = 2$ (goals per match).

In this situation the interval is a football match.

The Poisson distribution may also be used as an approximation to the binomial distribution, B(n, p), when n is large, p small and np is neither large nor very small. The mean is np and this is the value of the parameter λ.

The Poisson distributions for $\lambda = 1$, 2, and 3 are shown in the graphs of Fig. 98.

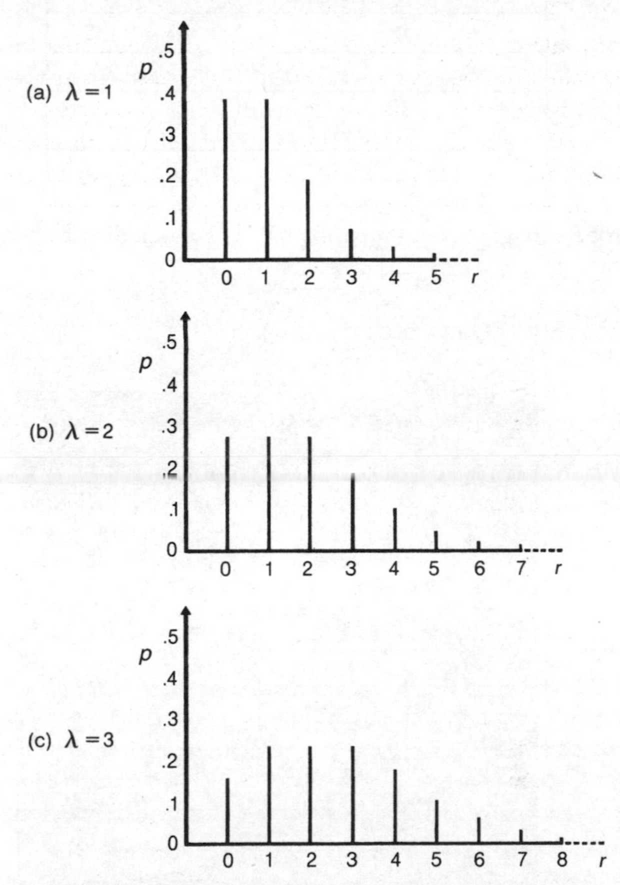

Fig. 98. **Poisson distribution.** (a) $\lambda = 1$ (b) $\lambda = 2$ (c) $\lambda = 3$.

Typical situations which can be modelled by the Poisson distribution are the number of hedgehogs killed on a particular section of road per day and the number of telephone calls coming into an exchange in a period of one minute.

Example: in a large maternity hospital there are, on average, 2 pairs of twins every week. Find the probabilities of 0, 1, 2, 3 sets of twins being born in any week.

The mean incidence of twins per week is 2 and so the parameter $\lambda = 2$.

No. of pairs of twins	Probability
0	$e^{-2} \times 1 = 0.135$
1	$e^{-2} \times 2 = 0.271$
2	$e^{-2} \times \frac{2^2}{2!} = 0.271$
3	$e^{-2} \times \frac{2^3}{3!} = 0.180$

The probability generating function for the Poisson distribution is:

$$\sum_r t^r \frac{e^{-\lambda}\lambda^r}{r!} = e^{\lambda(t-1)}$$

(See PROBABILITY GENERATING FUNCTION).

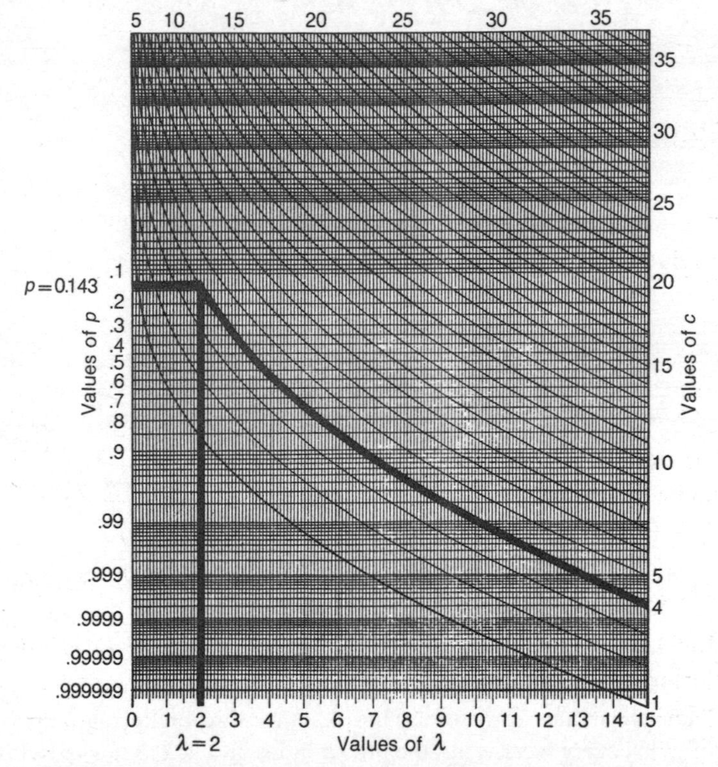

Fig. 99. **Poisson distribution.** Poisson probability chart; when $\lambda = 2$ and $c = 4$, $p\ (x \geqslant 4) = 0.143$.

Calculations using the Poisson distribution can be rather cumbersome if the probability required is that of 'less than', 'more than', 'at most' or 'at least n occurrences' where n is not particularly small. Some books of tables include Cumulative Poisson Probability tables to overcome this. Another approach involves the use of the *Poisson probability chart* (Fig. 99. Compare GRAPH). It allows one to find the probability p of at least c occurrences for any mean. Thus, in the example of the twins, the question 'What is the probability of at least 4 pairs of twins in any week?' could have been asked. The answer from the previous work is

$$1 - (0.135 + 0.271 + 0.271 + 0.180) = 0.143$$

This can be found on the chart by finding the value of p where the line $\lambda = 2$ and the curve $c = 4$ cross.

Poisson process the stochastic process that underlies the Poisson distribution.

polar graph see GRAPH.

population **1.** the complete set of values of a variable in a given situation (this is also called the *parent population*).

2. the individuals making up a group of humans or animals.

population pyramid, population profile a type of frequency chart used to display the population age structure of a country (or city, region, etc). The population figures are divided into age groups, and into male and female. The scale for the bars may be for the actual numbers or for their percentages.

There are three main types of age structure of a given population; each has a distinctive population profile. If children are defined as under 16 and the aged as 65 and over, a typical categorisation might be as follows (Fig. 100):

Progressive: over 45% children, and under 10% aged. This population will grow.

Regressive: under 30% children, and over 15% aged. This population is likely to diminish.

Stationary: 35-40% children, and 10% aged. This population will stay reasonably static.

An *intermediate* structure is one changing from one type to another.

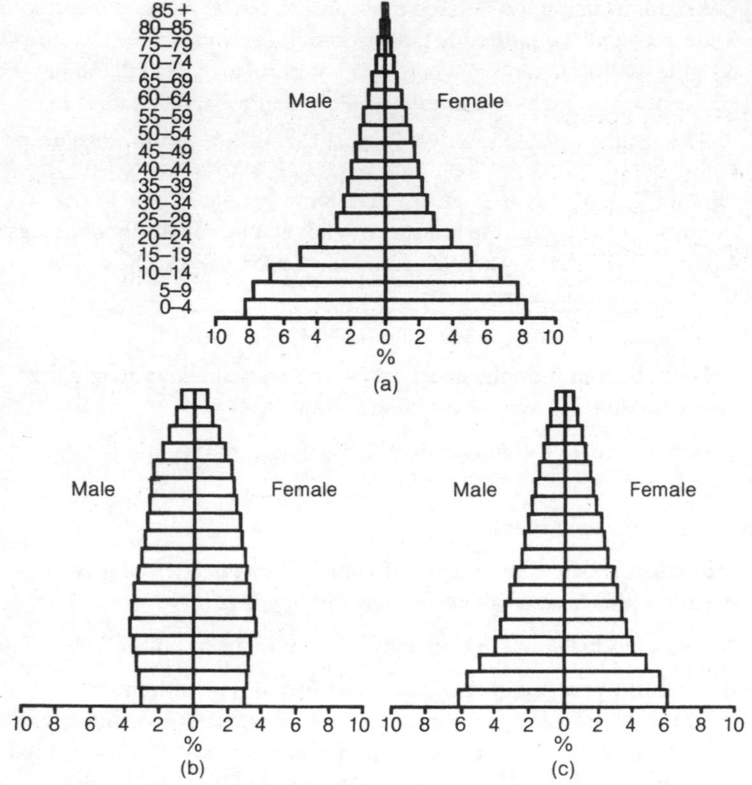

Fig. 100. **Population profile.** (a) Progressive. (b) Regressive. (c) Stationary.

posterior probability see PRIOR PROBABILITY.

power for a given statistical test, the function which gives the probability that the null hypothesis is rejected at any value of the parameter being tested. If a null hypothesis is true and it is being tested at the 5% significance level, there is a 5% probability of it being falsely rejected. Thus the value of the power is 0.05 when the value of the parameter being tested coincides with its true value. When a Type 2 error occurs a false null hypothesis is accepted, so for any particular parameter value, λ,

$$\textit{Power}\ (\lambda) = 1 - \textit{Probability of a Type 2 error}\ (\lambda).$$

The function which gives the probability of a Type 2 error is the

Operating Characteristic of the test and so

$$Power = 1 - Operating\ Characteristic.$$

See OPERATING CHARACTERISTIC.

principal component analysis a technique for finding relationships in the variables in multivariate data, with the aim of allowing the dimension of the data to be reduced without serious loss of information.

prior probability a prior probability is a measure of beliefs about a situation prior to doing any experiment at all; this is often based on subjective judgement. By contrast the *posterior probability* takes experimental data into account.

The prior probability of an event, A, with respect to another event, B, is the probability that event A occurs if it is not known whether event B has occurred or not. This is thus $P(A)$.

The *posterior probability*, for the same pair of events, is the probability that event A occurs if it is known that event B has occurred. This is written $P(A|B)$.

Example: two dice are thrown, first one then the other. Before either is thrown the probability of a total of at least 9, the prior probability, is $\frac{5}{18}$. However, if the first die is seen to land 6, the probability of a total of at least 9, the posterior probability, is $\frac{2}{3}$.

In this example,

Event A is "The total is 9"
Event B is "The first die is 6"

Event A occurs if the total is 9, 10, 11 or 12. The probability of this, $P(A)$, is given by:

$$P(A) = \frac{4}{36} + \frac{3}{36} + \frac{2}{36} + \frac{1}{36} = \frac{10}{36} = \frac{5}{18}$$

The event $A|B$ occurs if, following an earlier throw of 6 on the first die, the second die comes up 3, 4, 5 or 6. The probability $P(A|B)$ that the second die shows 3, 4, 5, or 6, is

$$P(A|B) = \frac{4}{6} = \frac{2}{3}$$

If the prior and posterior probabilities are equal, events A and B are independent.

probability a measure, on a scale of 0 to 1, of the likelihood of an event occurring. A probability of 0 means impossibility and 1 means certainty. Probability may be calculated from a theoretical distribution or estimated following observations.

For a discrete distribution in which all the outcomes are equally likely, probability is defined as:

$$\frac{\text{the number of required outcomes}}{\text{the total number of possible outcomes}}$$

Thus the probability of drawing a spade from a pack of 52 playing cards is $\frac{13}{52} = \frac{1}{4}$, because

The number of required outcomes = 13 (spades).

The total number of possible outcomes = 52 (cards).

For a continuous variable, the probability is the relevant area under the graph of its probability density function.

In the case shown in Fig. 101, the probability density function f(x) is

$$f(x) = \tfrac{3}{32}(4x - x^2).$$

The probability that x lies between 1 and 2 is the shaded area, and this is given by

$$\int_1^2 \tfrac{3}{32}(4x - x^2)\,dx.$$

$$= \tfrac{11}{32}.$$

This area represents

$$\frac{\text{the frequency of results between 1 and 2.}}{\text{the total frequency of all results}}$$

See also IMPOSSIBLE, BAYESIAN CONTROVERSY.

probability density function the function f(x) of a continuous random variable X, such that

$$\int_a^b f(x)\,dx$$

is the probability that X lies between a and b.

In the diagram (Fig. 101) the probability that X lies between 1 and 2 is given by the shaded area, and is $\frac{11}{32}$.

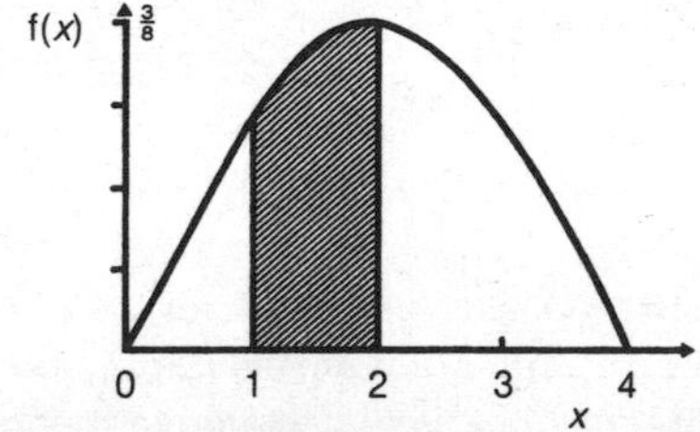

Fig. 101. **Probability density function.** $f(x) = \frac{3}{32}(4x - x^2)$

A probability density function, $f(x)$, must obey two conditions:

(i) the total probability for all possible values of X is 1.

$$\int_{\text{all } X} f(x)\,dx = 1$$

(ii) the probability can never be negative:

$$f(x) \geq 0 \text{ for all } x.$$

The mean (or expectation) and variance of X are given by:

Mean $\quad E(X) = \int_{\text{all } X} x f(x)\,dx$ (See EXPECTATION).

Variance $\quad \text{Var}(X) = E\left[(X-\mu)^2\right] = \int_{\text{all } X} (x-\mu)^2 f(x)\,dx$

At the mode of the distribution, $f'(x) = 0$, where f' is $\frac{df}{dx}$.

The median of the distribution is the value of x such that

$$\int_{-\infty}^{x} f(x)\,dx = \int_{x}^{\infty} f(x)\,dx = 0.5$$

Some commonly used probability density functions are given in the table below.

Distribution	p.d.f., $f(x)$
Normal, $N(\mu,\sigma^2)$	$\frac{1}{\sigma\sqrt{2\pi}} e^{-\frac{1}{2}\left(\frac{x-\mu}{\sigma}\right)^2}$
Standardised Normal, N(0,1)	$\frac{1}{\sqrt{2\pi}} e^{-\frac{1}{2}z^2}$ where $z = \frac{x-\mu}{\sigma}$
Continuous uniform on $[a,b]$	$\frac{1}{b-a}$
Exponential	$\lambda e^{-\lambda x}$

For discrete distributions, the equivalent of probability density function is probability distribution.

probability distribution, probability function or **probability mass function** the distribution of the probabilities of the different values of a discrete random variable, X. Notations $p(x_i)$, $P(X = r)$ for integer values of r.

Example: the probability distribution of the number of girls in families of 3 children, is as follows:

x_i (no. of girls)	0	1	2	3
Probability $p(x_i)$	$\frac{1}{8}$	$\frac{3}{8}$	$\frac{3}{8}$	$\frac{1}{8}$

The probability distribution of a random variable X taking values x_1, x_2, x_3,...x_n, with probabilities $p(x_1)$, $p(x_2)$, ...$p(x_n)$, must satisfy two conditions:

(i) The sum of all the probabilities is 1;

$$\sum_{i=1}^{n} p(x_i) = 1.$$

(ii) No probability can be negative;

$$p(x_i) \geq 0 \text{ all } i.$$

The mean (or expectation) and variance of X are given by:

Mean $\quad \mathrm{E}(X) = \sum_{i=1}^{n} x_i \mathrm{p}(x_i)$ $\quad$ (See EXPECTATION.)

Variance $\quad \mathrm{Var}(X) = \mathrm{E}\left[(X-\mu)^2\right] = \sum_{i=1}^{n} (x_i-\mu)^2 \mathrm{p}(x_i)$

Some commonly used probability distribution functions are given in the table below.

Distribution	Function, $P(X = r)$
Binomial, $B(n, p)$	$\binom{n}{r} q^{n-r} p^r$ where $q = 1 - p$
Poisson (λ)	$e^{-\lambda} \frac{\lambda^r}{r!}$
Discrete Uniform	$\frac{1}{n-m+1}$ for $r = m,\ m+1,\ \ldots n$
Geometric	$q^{r-1} p$ where $q = 1 - p$
Negative binomial	$\binom{r-1}{n-1} q^{r-n} p^n$ where $q = 1 - p$ for $r = n,\ n+1,\ \ldots$

For continuous distributions, the equivalent of probability distribution is probability density function.

probability function see PROBABILITY DISTRIBUTION.

probability generating function an expression which, when expanded, has terms whose coefficients are the probabilities of the different numbers of occurrences of an event. Notation $G(t)$.

As an example, the expression

$$(\tfrac{1}{2} + \tfrac{1}{2}t)^3$$

can be expanded to give

$$\tfrac{1}{8} + \tfrac{3}{8}t + \tfrac{3}{8}t^2 + \tfrac{1}{8}t^3,$$

representing the probabilities of $\frac{1}{8}, \frac{3}{8}, \frac{3}{8}$, and $\frac{1}{8}$ of 0, 1, 2 and 3 tails respectively when a coin is tossed three times.

In general the probability generating function of a discrete random variable X is given by $G(t) = E(t^X)$ and this can, alternatively, be written as $G(t) = \sum_r t^r P(X = r)$.

For a probability distribution with probability generating function $G(t)$

$$G(1) = 1$$
$$E(X) = \mu = G'(1)$$
$$\mathrm{Var}(X) = G''(1) + \mu - \mu^2$$

where $G'(1)$ means the value of $\frac{dG}{dt}$ when $t = 1$, etc.

For independent random variables, X and Y,

$$G_{X+Y}(t) = G_X(t) \times G_Y(t).$$

The probability generating functions for some common distributions are as follows.

Distribution	Function, $P(X = r)$	p.g.f., $G(t)$
Binomial, $B(n, p)$	$\binom{n}{r} q^{n-r} p^r$ where $q = 1 - p$	$(q + pt)^n$
Poisson (λ)	$e^{-\lambda} \dfrac{\lambda^r}{r!}$	$e^{\lambda(t-1)}$
Geometric	$q^{r-1} p$ where $q = 1 - p$	$\dfrac{pt}{1 - qt}$
Negative binomial	$\binom{r-1}{n-1} q^{r-n} p^n$ where $q = 1 - p$	$\left(\dfrac{pt}{1 - qt}\right)^n$

probability mass function see PROBABILITY DISTRIBUTION.

probit a function of a probability. The probit of p is given by

$$\mathrm{probit}(p) = \Phi^{-1}(p) + 5$$

where $\Phi^{-1}(p)$ is the inverse of the cumulative Normal probability $\Phi(p)$.

Example: find the probit of $p = 0.95$.

$$\mathrm{probit}(0.95) = \Phi^{-1}(0.95) + 5$$

Table 2 shows that $\Phi(1.645) = 0.95$ and so

$$\text{probit}(0.95) = 1.645+5 = 6.645.$$

The probit has been widely used in modelling effective doses of medicines, insecticides, etc. The word *normit* is sometimes used for $\Phi^{-1}(p)$. The purpose of adding 5 to it is to be almost certain to avoid negative values.

process average in quality control, the proportion of defective items produced by a machine or manufacturing process when it is running normally.

producer's risk in quality control, the producer's risk is the risk that a batch with proportion of defectives equal to the process average (i.e. an average batch) will be rejected by the inspection scheme, or the probability that this will happen. Historical note: producer's risk was at one time a name for Type 1 error.

product-moment correlation see CORRELATION COEFFICIENT.

proportional allocation see STRATIFIED SAMPLING.

proportions distribution the distribution of the proportion of successes in a set of Bernoulli trials. If each trial has probability p of success and $q = (1 - p)$ of failure, this distribution has

$$Mean = p, \quad Standard\ deviation = \sqrt{\frac{pq}{n}}.$$

This is just the binomial distribution $B(n, p)$ scaled by dividing by n. The distribution $B(n, p)$ has mean np and standard deviation $\sqrt{npq}$ and dividing these by n gives the mean and standard deviation for the proportions distribution, as stated above.

q

Q-Q plot a plot of quantiles of the empirical distribution of a data set against the corresponding quantiles of a theoretical distribution proposed as a model, widely used as a model checking diagnostic.

quadrat an area, usually of vegetation, randomly selected for study. It is normally square in shape (hence the name). The ideal size of a quadrat is the smallest size that contains the same number of species as would be contained in a larger one. In a *point quadrat*, sampling is carried out at the points of a square grid covering the quadrat.

When several quadrats are placed in a row, the area formed is called a *transect* (Fig. 102).

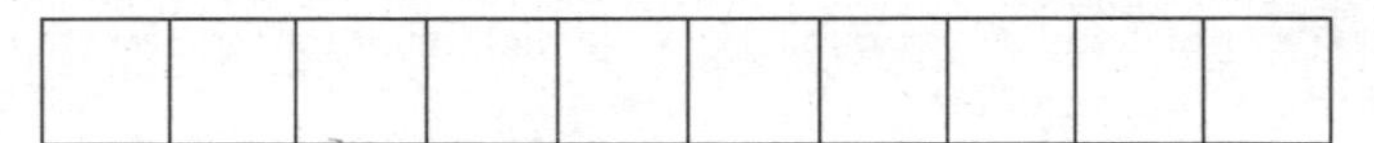

Fig. 102. **Quadrat.** Transect comprising 10 quadrats.

qualitative data data which are classified by type. Thus data on how hospital patients feel after a course of treatment can be grouped qualitatively as:

Much better   Better   The same   Worse   Much worse

If however, their numbers of hours sleep are recorded, these data are *quantitative data*, being expressed numerically.

quality control a general name for the processes used to ensure the quality of a product. Many of these processes involve taking and interpreting samples. The product is often a manufactured item but the general ideas are also used in many service industries.

quantile a general name for the values of a variable which divide its distribution into equal groups. A set of numbers may be ranked according to size.

Thus the numbers 6, 8, 6, 3, 9, 5, 6, 5, 11, 4, 12, 10, 9, 2, 2, 5, 5, 7, 3, 11, 6 can be ranked as:

No.	2	2	3	3	4	5	5	5	5	6	6	6	6	7	8	9	9	10	11	12	12
Rank	1	2	3	4	5	6	7	8	9	10	11	12	13	14	15	16	17	18	19	20	21

The ranking used to find quantiles is slightly different from that used at other times; when a tie occurs, the values are ranked one above the other, rather than equal.

A set of ranked numbers is divided into four equal groups by the quartiles, into six equal groups when the divisions occurs at the sextiles, into eight with divisions at the octiles, ten at the deciles, and a hundred the percentiles. Quartiles, sextiles, octiles, deciles and percentiles are all examples of quantiles.

The median or second quartile has the middle position. In the set of numbers given above, the median is the number ranked 11, namely 6. For n numbers, the median is that with rank $\frac{1}{2}(n+1)$ in the ranked order. If n is even there is no actual middle number and the arithmetic mean of the two numbers on either side of the middle is taken to be the median. See QUARTILE.

quantitative data see QUALITATIVE DATA.

quartile the value of a variable below which one quarter (1st or lower quartile) or three quarters (3rd or upper quartile) of a distribution lie. The median is the 2nd quartile. Notation Q_1, Q_2, Q_3.

Difficulties arise when quartiles are used for small data sets because there is no single agreed convention for assigning values to the quartiles.

There are two different cases to be considered, according to whether the size of the data set is even or odd. In the examples which follow these cases are illustrated for an even set of size 12 and an odd set of size 13.

Case 1. An even sized data set: {3, 4, 4, 5, 6, 7, 9, 11, 11, 12, 14, 15}

No ambiguity arises in this case. There is no single middle value so the median is the value half way between the two either side of the middle, $\frac{1}{2}(7+9) = 8$.

The data can be partitioned into two equally sized subsets, each with 6 members: {3, 4, 4, 5, 6, 7} and {9, 11, 11, 12, 14, 15}.

The lower quartile is the median of the first of these two subsets, $\frac{1}{2}(4+5) = 4.5$, and the upper quartile is the median of the other subset, $\frac{1}{2}(11+12) = 11.5$.

The lower quartile, median and upper quartile partition the data set into four equally sized subsets: {3, 4, 4}, {5, 6, 7}, {9, 11, 11} and {12, 14, 15}.

Case 2. An odd sized data set: {3, 4, 4, 5, 6, 7, 9, 11, 11, 12, 14, 15, 17}

The median of this set is 9. However there is no agreement as to whether the median value of 9 should, or should not, be included in the two subsets into which the data are now partitioned.

If the median value of 9 is not included, the two subsets are {3, 4, 4, 5, 6, 7} and {11, 11, 12, 14, 15, 17}. The quartiles are then said to be the medians of these two subsets, 4.5 and 13.

If, however, the median value of 9 is included, the two subsets are {3, 4, 4, 5, 6, 7, 9} and {9, 11, 11, 12, 14, 15, 17}. The quartiles are now said to be 5 and 12.

The situation is further complicated by a commonly quoted "rule", that the lower quartile, median and upper quartile are the values ranked at $\frac{1}{4}(n+1)$, $\frac{1}{2}(n+1)$, $\frac{3}{4}(n+1)$ respectively, where n is the size of the data set. In some cases these give the same values as the procedures stated above and in other cases they do not.

One way of avoiding the problem is not to use quartiles for small data sets. If, for example, the upper quartile of the heights of 13 year old girls in the UK is being sought, it makes no practical difference which method is being used.

However, in Exploratory Data Analysis quartiles are routinely used with small data sets. To avoid any possible ambiguity, Tukey (who developed this branch of statistics) introduced the word *hinge* as a replacement for *quartile* with the proviso that in calculating the hinge of an odd sized data set, the median is included in each of the two subsets whose medians give the hinges (or "quartiles"). Thus in Case 2, above, the hinges are 5 and 12. The hinges corresponding to the lower and upper quartiles are denoted by H_1 and H_2, repectively, and the quantity $H_2 - H_1$ is called the H-spread. The H-spread is comparable to inter-quartile range. See also INTER-QUARTILE RANGE, DISPERSION, EXPLORATORY DATA ANALYSIS, QUANTILE.

quartile coefficient of dispersion see ABSOLUTE MEASURE OF DISPERSION.

quartile coefficient of skewness see SKEW.

quartile deviation see INTERQUARTILE RANGE.

questionnaire a set of questions used to collect information from people.

quota sampling a sampling method in which one of more interviewers are used, each being given instructions about the section of the population which is his or her responsibility. The actual choice of interviewees is left to the interviewer. Each of them might, for example, be told to select 20 adult women, 20 adult men, 10 teenage girls and 10 younger children from their home town. Quota sampling is often used in market and social surveys.

Quota sampling is non-random with respect to the population being surveyed; this leads to substantial complications in the statistical analysis of the survey results. Nevertheless, quota sampling is very widely used. It is often the only practical way to carry out a survey.

r

random error the difference between an observed data value and its expected value due to the random variability inherent in the situation under study. See ERROR.

randomisation the process of ensuring that, when possible, the elements in a statistical experiment are carried out in a random order.

Randomisation is one of the key principles in designing an experiment. It is a safeguard against systematic errors.

Example: an experiment is to be carried out to investigate the ability of a person to estimate times of 5 seconds, 10, seconds, ... , 30 seconds. The times are taken in order and the subject's estimates have greater relative error for the longer time intervals. This could be because the subject gets more bored during the longer time intervals and wants to get them over, or it could be that the subject gets bored with the whole experiment. If this possibility is not considered until after the experiment is completed, there may be nothing that can be done to decide which is the correct explanation. If the subject is no longer around, it is impossible to find out whether the effect was due to the longer times or to the length of the experiment. The two effects are confounded.

Instead, the order in which the person estimates the times should be randomised. If the relatively large errors still occur with the longer time intervals, then it may well be that the effect is due to boredom during the longer intervals. If, on the other hand, large errors occur later in the experiment, the effect may be caused by boredom with the whole experiment.

randomised blocks design see EXPERIMENTAL DESIGN.

randomised response technique a technique for obtaining information which might be damaging or incriminating to individuals.

Example: A social worker is investigating domestic violence. She wants to know what percentage of women in her town suffer violence from their male partners. However she believes that if she asks the partner directly some who are violent will deny it.

Instead she gives a sample of 100 male partners a sheet of paper one Friday with the following instructions.

Toss a coin
If it comes heads, answer question A.
If it come tails, answer question B.
A. Is it Friday today?
B. Are you ever violent towards your partner?

Yes/No

60 reply "Yes"; the other 40 reply "No".

There is no way that the social worker can tell whether a particular "Yes" is in response to question A or to question B. So confidentiality is not broken. It is however, possible for her to use the response to estimate the proportion of violent male partners in the town.

Out of the 60 "Yes" responses, 50 may be expected to be answering question A, since the probability of the coin coming heads is $\frac{1}{2}$ and 100 people took part in the experiment.

That leaves 10 violent partners out of the 50 who got tails, and answered question B. The social worker concludes that approximately 20% of the partners are violent.

Considerable caution has to be taken when interpreting the results of the method. It is quite possible that 55 of the partners in the above example got heads when they tossed the coin rather than 50. In that case only 5 out of 45 answered Yes to question B, about 11% and not the calculated 20%. A very large sample is needed to obtain reliable results. In addition it is assumed that, in these circumstances, everyone answers truthfully.

random numbers numbers generated by a random process in which each number is independent of all previous numbers generated. The numbers are drawn from a population with a known distribution. Commonly it is a uniform distribution and in that case each number has an equal chance of being generated.

Such random numbers are given in tables like those in Table 17.

Each of these numbers is selected at random from the digits 0, 1, 2, 3, 4, 5, 6, 7, 8 and 9. Computers and some calculators have a random number generating facility.

Random numbers are used in the design and implementation of random sampling schemes. They are also used in simulation; for example, a flight simulator for training pilots uses random numbers to generate gusts and changes of direction to be imposed on a steady basic wind.

Random numbers may also be generated from other distributions, for example the Normal distribution, N(0, 1).

random sampling see SAMPLING.

random selection selection in which all items have a known, non-zero probability of being selected. See, for example, SIMPLE RANDOM SAMPLING.

random walk a random process in which a sequence of discrete steps of fixed length is described in terms of the movement of a particle.

In a one-dimensional random walk, the state of the process is described as a position on a straight line. After starting at 0 (say), the first step takes the particle to either +1 or –1; after two steps it is at +2, 0 or –2, and so on.

A two-dimensional random walk can be represented either by a set of steps along a unit grid or lattice, for example a sheet of squared paper. In a different representation, the steps are of fixed length but at random angles. (See Fig. 103.)

The particle may represent any suitable quantity, for example an amount of money.

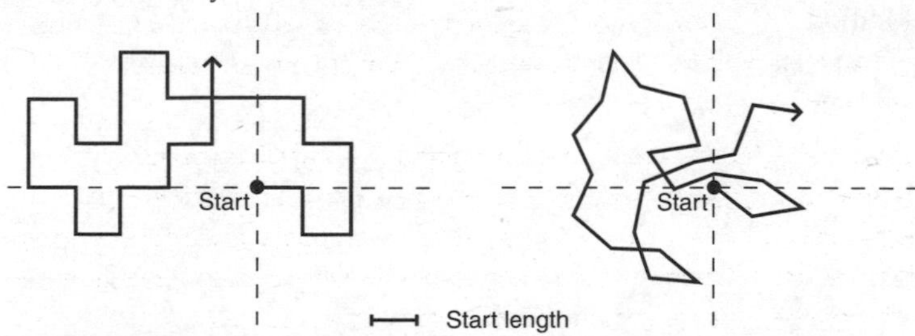

Fig. 103. **Random walk.** Two different representations of random walks in 2-dimensions.

random variable a variable which takes values in a certain range with probabilities that can be specified by a probability distribution or probability density function.

When a coin is tossed, the outcome is a head or a tail. This is not a random variable because it is not a number. If, however, the result is expressed as the number of tails, it is a random variable which can take the values 0 and 1, each with probability $\frac{1}{2}$. Alternatively the results may be coded, for example 1 for a head and 2 for a tail, thereby creating a random variable.

By convention, upper case letters are used to denote the "names" of random variables, lower case for particular values of them. So, for example, X could be the random variable "The number of occupants in cars coming into Plymouth between 8 am and 9 am". X can take values 1, 2, 3, 4, … . If, in an experiment, a sample of 20 cars is to be taken, the numbers of occupants of those cars would be denoted by $x_1, x_2, \ldots x_{20}$.

range the difference between the largest and the smallest values of a data set.

The range of 47, 52, 58, 63, 65 is 65-47 = 18.

The standard deviation of a Normally distributed population may be estimated from the range of a sample, using Table 10. See DISPERSION.

rank the position of an item within a data set, when all the items are arranged in order according to size. The ranking may be done in ascending or descending order.

When two or more items are equal, they are said to be *tied*. There are two different conventions for assigning ranks to tied data and they are illustrated in the example below.

Example: rank the numbers 47, 50, 49, 65, 68, 49, 49, 65.

Method 1

Number	*Rank*	
47	1	
49	3	The mean of 2, 3 and 4 is 3.
49	3	
49	3	
50	5	
65	$6\frac{1}{2}$	The mean of 6 and 7 is $6\frac{1}{2}$
65	$6\frac{1}{2}$	
68	8	

In this method the tied items are all given the same rank, that of the mean of the ranks they would have had if they had been slightly different.

Many non-parametric tests are based on rank, for example Kendall's and Spearman's rank correlation tests, the Wilcoxon tests and the Kruskal Wallis one-way analysis of variance and this method is used for all of them.

Method 2

Number	*Rank*
47	1
49	2
49	3
49	4
50	5
65	6
65	7
68	8

In this method the tied items are given the ranks they would have had if they had been slightly different.

This method is used in finding quantiles and hinges.

Notice that neither of these methods is that used in everyday life. If, for example, four children scored 5.7, 5.5, 5.5 and 5.2 in a skating contest, they would usually be ranked 1, 2=, 2= and 4 respectively.

For most tests involving ranks it does not matter if the rank starts with the highest or lowest number as rank 1, so long as the test is consistent throughout. This is not true, however, of the Wilcoxon matched-pairs signed-rank test, where the smallest absolute difference must count as rank 1.

rank correlation correlation of bivariate data by rank, rather than by numerical value. This can be measured using Spearman's or Kendall's rank correlation coefficient. See CORRELATION COEFFICIENT.

ratio scale a scale of measurement in which equal differences between points correspond to equal distances on the scale, and there is a true zero.

An example of a ratio scale is length. The difference between 2 m and 2.50 m is the same as that between 10.25 m and 10.75 m.

A length of zero means no length. Consequently it is possible to speak of ratios of lengths. A length of 6 m is twice as long as one of 3 m.

rectangular distribution or continuous uniform distribution a continuous distribution which is constant over a certain range of values, and zero elsewhere. This is illustrated in Fig. 104 where $a \leq x \leq b$; the graph has a rectangular shape.

$$\text{Mean} \quad \mu = \frac{a+b}{2} \qquad \text{Variance} \quad \sigma^2 = \tfrac{1}{12}(b-a)^2$$

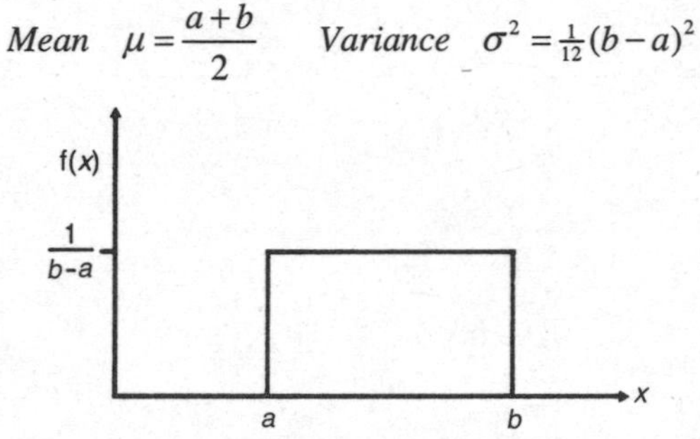

Fig. 104. **Rectangular distribution** or **continuous uniform distribution.** In this example, $f(x) = 1/(b - a)$, where $a \leq x \leq b$ $f(x)=0$ elsewhere. Thus the area under the graph is 1.

An alternative name for the rectangular distribution is the continuous uniform distribution. See UNIFORM DISTRIBUTION.

reflecting barrier a state for a random walk which, when crossed in a given direction (say "downwards"), holds the particle until an "upwards" jump occurs, whereupon the random walk resumes.

Example: the amount of water in a reservoir, in suitable units, at a given time (say midday) on day n ($n = 1, 2, \ldots$) is represented by X_n. X_n can be modelled as undertaking a random walk with possible values 0, 1, 2, ..., c, where $X_n = 0$ means that the reservoir is empty, and c is the capacity of the reservoir. The value of X_n varies from day to day depending on inflow (rainfall, rivers flowing into the reservoir) and outflow (the amount of water taken from the reservoir each day). $X_n = 0$ is a reflecting barrier: if the reservoir is empty and an attempt is made to take water out of it, it stays empty until some water has accumulated through inflow. $X_n = c$ is also a reflecting barrier: the reservoir cannot become more than full (any excess water would simply be lost, perhaps by overflowing a dam); while inflow (rainfall) continues at a higher rate than outflow, it remains full, until that situation no longer applies.

regression line a linear relationship representing a regression model for two variables. It can be drawn as a straight line on a scatter diagram. It is quite common practice, particularly in the initial stages of an investigation, to do this informally by drawing a line of best fit through a set of data points by eye.

There are two different circumstances in which regression lines are often drawn.

1. The case where Y is a random variable but x is a non-random variable (or, the x-values are regarded as given values from an underlying random variable). This often arises in designed experiments.

Example: a chemical reaction is run at several different temperatures (x_1, x_2, ...) and the yield (Y) is measured in each case. The x-values, having been deliberately chosen by the experimenter, can be regarded as being fixed values essentially without experimental error; so x is a non-random variable. However, the yield is expected to vary if the experiment is repeated several times at any particular x-value (due, for example, to changes in ambient conditions and the quality of the raw materials). So Y is a random variable.

In this case, it is appropriate to use the method of least squares to find the regression line. Its formula works out to be

$$y-\overline{y}=\frac{S_{xy}}{S_{xx}}(x-\overline{x}).$$

Example: find the regression lines through the points (2, 3), (4, 5), (6, 6), (5, 8) and (8, 13).

x	y	$x-\overline{x}$	$y-\overline{y}$	$(x-\overline{x})(y-\overline{y})$	$(x-\overline{x})^2$	$(y-\overline{y})^2$
2	3	-3	-4	12	9	16
4	5	-1	-2	2	1	4
6	6	1	-1	-1	1	1
5	8	0	1	0	0	1
8	13	3	6	18	9	36
25	35			35	20	58

$\overline{x}=\frac{25}{5}=5$ $\overline{y}=\frac{35}{5}=7$

$$S_{xy} = \sum_{i=1}^{n} (x_i - \overline{x})(y_i - \overline{y}) = 35$$

$$S_{xx} = \sum_{i=1}^{n} (x_i - \overline{x})^2 = 20$$

Thus the least squares regression line, y on x, is

$$y - 7 = \frac{35}{20}(x - 5)$$

$$y = 1.75x - 1.75$$

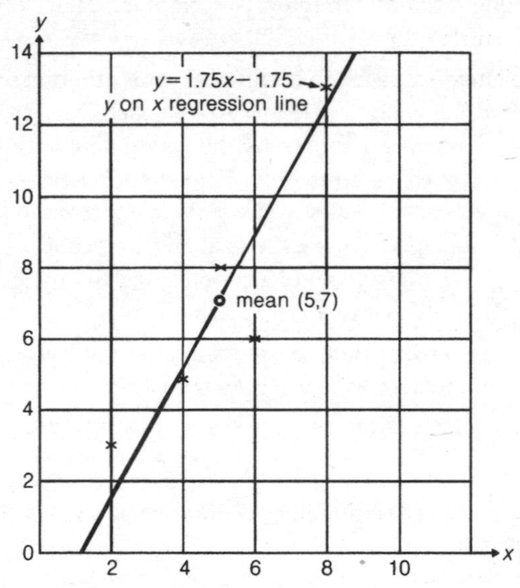

Fig. 105. **Regression line.** The y on x regression line, and the mean for the points (2,3), (4.5), (6.6), (5,8) and (8,13).

2. The case where both X and Y are random variables.

Example: in an experiment the heights and weights of a random sample of people are measured. The experimenter did not deliberately choose people of various weights and measure their heights (or vice versa); instead, the (x, y) values were simply a random sample from a bivariate random variable (X, Y).

It is still sensible to seek to draw regression lines, which now model the expected value of one variable for given values of the other. There are now two regression lines, one for Y on x and the other for X on y, and in general these are different. Theoretically, they should be found by the method of maximum likelihood, which requires knowlege about the statistical behaviour of the bivariate random variable (X, Y). However, fortunately, the most commonly occurring and useful case turns out to be very easy. This is where (X, Y) has the bivariate Normal distribution; it is very often a good model. It works out that the same formulae apply in this case as would have been obtained by using the method of least squares (see case (1) above) separately to find the Y-on-x line and the X-on-y line. See LEAST SQUARES, METHOD OF, BIVARIATE NORMAL DISTRIBUTION.

regression model *n* any model relating a random variable Y to one or more explanatory variables x. If there is more than one explanatory variable, they are often written as a vector

$$\mathrm{x} = \begin{pmatrix} x_1 & x_2 & \ldots & x_n \end{pmatrix}.$$

A regression model is a type of linear model. Simple linear regression is the simplest such situation and the idea of a regression line arises from it.

relative frequency the ratio of the frequency of a particular outcome to the total number of experiments.

$$\text{Relative frequency} = \frac{\text{frequency of a particular outcome}}{\text{total number of experiments}}$$

Relative frequency is often used as an estimate of probability.

reliability coefficient a measure of consistency obtained by calculating the correlation coefficient between two repetitions of the same experiment. In the *split-half method*, the correlation is carried out on the results of the two halves of one experiment.

replication the process of repeating an experiment with the circumstances kept the same, as far as possible.

Replication is one of the key principles in the design of experiments. The purpose is to assess the experimental error, and hence the reliability of any conclusions drawn from the results.

Example: an engineer wishes to investigate how the length of a spring depends on the weight which is hung from it. From a pilot experiment, he has decided that a suitable range of weights is 10 g to 100 g. He randomises the order of the weights and records the extensions of the spring, and then repeats this process several times to give the replicates. When the results are plotted on a scatter graph a linear model seems reasonable and so he finds this, using least squares regression. He then uses the model to predict the extension of the spring for a given weight. The variability of the extensions measured for each value of the weight gives him some idea about the reliability of his model. If the points are very close together then the forecasts from his model should be very good.

The replicates also enable the engineer to model the variability in his measurements. He could see whether a Normal model seemed reasonable. He can also see whether the variability in the extensions remains the same for different weights, or whether it seems to depend on the weight used. He might then be able to produce confidence intervals for his estimates.

residual the difference, $y - \hat{y}$ between the observed value, y, of a variable and the corresponding value, $\hat{y}$, obtained by fitting it to a model.

In the case of the least squares y on x regression line, a residual is the vertical distance from a point to the regression line when plotted on a scatter diagram. See Fig. 106.

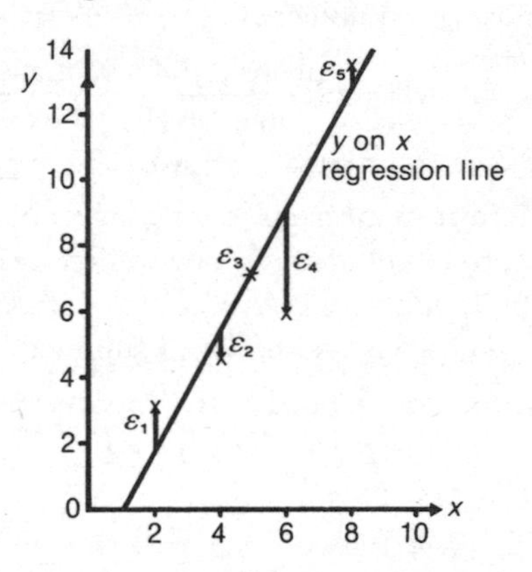

Fig. 106. **Residual.** Residuals $\varepsilon_1, \varepsilon_2, \ldots \varepsilon_5$, from the y on x regression line.

residual sum of squares see ANALYSIS OF VARIANCE.

resistant line a line of best fit for bivariate data plotted on a scatter diagram.

The resistant line is an alternative to the least squares regression line. It has the advantage that, as its name suggests, it is resistant to the effects of outliers. This is particularly useful when there are clear outliers in the data set but no obvious justification for removing them.

In the technique devised by Tukey, the data are divided into three groups of (almost) equal numbers on the basis of the x-values. A data set of size 30 would be split into three equal groups of size 10; with a data set of size 31, the extra value would be added to the middle group and with a data set of size 32, the two extra values would be placed in the two outside groups. However, over-riding these groupings is the principle that observations with the same x-value should go in the same group.

The equation of the resistant line is given by $y = mx + c$ where m and c are calculated as follows.

- Each of the three groups is assigned a summary point with coordinates (x-median, y-median).
- The gradient, m, of the resistant line is the gradient of the line joining the summary points of the two outside groups.
- The y-intercept, c, of the resistant line is given by the mean of the y-intercepts of the three lines of gradient m going through the three summary points. This is equivalent to putting a line through the two outside summary points and then moving it one third of the way towards the middle summary point.

Example: find the resistant line for the following points:

(17, 25) (26, 31) (19, 30) (15, 12) (30, 45) (30, 29) (24, 26)
(18, 2) (34, 36) (34, 46) (24, 40) (16, 12) (15, 20) (28, 36)
(14, 16) (27, 39)

There are 16 points and so they are placed in groups of size 5, 6 and 5 as follows.

Group						Summary point
(14, 16)	(15, 12)	(15, 20)	(16, 12)	(17, 25)		(15, 16)
(18, 2)	(19, 30)	(24, 40)	(24, 26)	(26, 32)	(27, 39)	(24, 31)
(28, 36)	(30, 29)	(30, 45)	(34, 36)	(34, 46)		(30, 36)

$$m = \frac{36-16}{30-15} = \frac{4}{3}$$

The lines with gradient $\frac{4}{3}$ through the three summary points are:

$y-16 = \frac{4}{3}(x-15)$ or $y = \frac{4}{3}x-4$. This line has intercept -4.

$y-31 = \frac{4}{3}(x-24)$ or $y = \frac{4}{3}x-1$. This line has intercept -1.

$y-36 = \frac{4}{3}(x-30)$ or $y = \frac{4}{3}x-4$. This line has intercept -4.

The mean of the three intercepts is $\frac{-4-1-4}{3} = -3$.

The resistant line is $y = \frac{4}{3}x-3$.

Notice the y-value in the point (18,2). It is an outlier but it does not influence the resistant line.

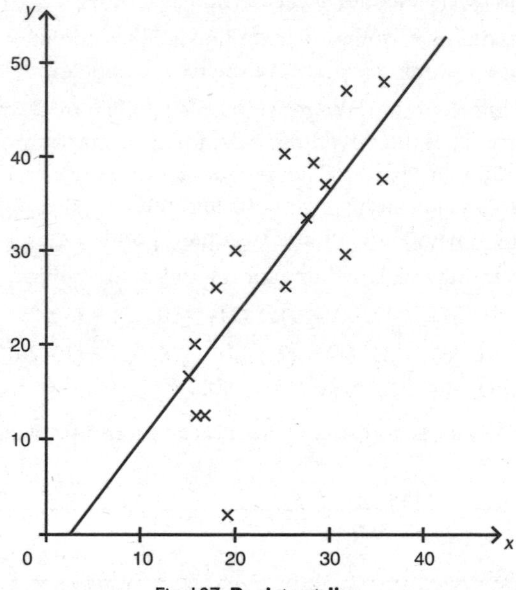

Fig. 107. **Resistant line.**

There are other related techniques.

In *Theil's incomplete method*, the line is calculated as $y = mx+c$ where m and c are calculated as follows. The sample is divided into two equal sets $(x_1, y_1), (x_2, y_2) \dots (x_N, y_N)$ and $(x_{N+1}, y_{N+1}), (x_{N+2}, y_{N+2}) \dots (x_{2N}, y_{2N})$ where the x values are in ascending order. (The middle point is omitted if the size is odd.) The gradients of the lines joining (x_1, y_1) to (x_{N+1}, y_{N+1}), (x_2, y_2) to (x_{N+2}, y_{N+2}), and so on are found. The median of these gradients has value m. Similarly c is the median of the intercepts of these lines.

In *Theil's complete method*, the quantity m is calculated as the median for all the possible pairings of points.

response variable see DEPENDENT VARIABLE.

retrospective analysis analysis done some time after the data have been collected, instead of (or as well as) analysing them immediately.

Example: a manufacturing process is being monitored on-line using Shewhart charts. The operator should respond immediately to a value which is outside the action limits. Although CuSum charts may also be used on-line, they are harder to interpret, and so these are often used for retrospective analysis. Thus, if a process has been found to have gone wrong, a CuSum chart may be produced. From this it is often possible to find the time at which the mean changed, and then to investigate whether an assignable cause can be found for the problem. See CONTROL CHARTS.

root mean square deviation the square root of the mean of the squares of the deviations of a set of numbers from their mean. The root mean square deviation of $x_1, x_2, \dots x_n$ is given by:

$$\sqrt{\frac{\sum_{i=1}^{n}(x_i - \overline{x})^2}{n}}$$

Example: the mean of 5, 4, 4, 3, 2, and 0 is 3. The root mean squared deviation is

$$\sqrt{\frac{2^2 + 1^2 + 1^2 + 0^2 + (-1)^2 + (-3)^2}{6}} = 1.63$$

rounding replacing a number by the nearest number to a given level of accuracy. There are several ways in which the level of accuracy may be presented, such as to a given number of significant figures, to a given number of decimal places, to the nearest whole number, to the nearest hundred, thousand, etc.

When a number is rounded, the usual procedure is as follows.

- Look at the figure immediately to the right of the last one required.
- If this figure is 5 or more, increase that to the left of it by 1 (rounded up); otherwise leave it unaltered (rounded down).
- Replace all digits to the right of the one required by 0; if, however, they are to the right of the decimal point omit them.

Thus to one decimal place: 23.152 is rounded up to 23.2,
23.148 is rounded down to 23.1.

Notice that it is wrong to take two or more steps when rounding; for example, going from 23.148 to 23.15 and then to 23.2 gives the wrong answer.

In the case of a number like 23.152, there is no doubt that it should round up, because it is nearer 23.2 than 23.1. In the case of 23.15, however, the decision whether to round up or down is arbitrary. The usual procedure is to round up as described above.

If, however, this procedure is applied to statistical data it can lead to systematic error.

For example, adding 23.15, 23.25, 23.35 and 23.45 gives a total of 93.20. However, if each of these figures is rounded to one decimal place, and then added, an error is produced:

$$23.2 + 23.3 + 23.4 + 23.5 = 93.4$$

To overcome this, an alternative procedure is sometimes used in this type of situation. A number is rounded up if the digit to the left of the 5 is odd, down if it is even. Thus,

23.15 is rounded up, 23.2
23.25 is rounded down, 23.2
23.35 is rounded up, 23.4
23.45 is rounded down, 23.4

When these new rounded figures are added, the total is now correct at 93.2. This procedure is sometimes called *statistical rounding*.

Other types of rounding are in common everyday use. For example: ages (in years) are rounded down, weights (and therefore prices) of postal packets are always rounded up.

running medians a method for smoothing a time series.

This method is similar to moving averages but uses the medians of the data rather than the means. Suitable medians are plotted as a time series from which seasonal or cyclic variation has been removed, allowing the underlying trend to be seen. The advantage of using medians over using means (i.e. moving averages) is that extreme points have a reduced effect. The method is resistant to outliers.

The first step is to look for any possible periodicity in the data and to choose the number of points so that each median covers one complete cycle of variation. So, for example, with monthly data showing annual variation it would be usual to use a 12-point running median, with quarterly data a 4-point running median, etc.

Example: a new daily local newspaper is launched. The sales (in hundreds) for the first three weeks are as follows.

	Mon	Tues	Wed	Thurs	Fri	Sat	Sun
Week 1	5.5	5.6	5.8	5.2	6.1	6.6	2.9
Week 2	6.2	6.2	6.6	5.9	6.9	7.2	3.5
Week 3	7.0	6.8	7.3	6.5	7.7	8.0	4.1

There is clearly a cyclic pattern to these figures. With the exception of Thursdays, they rise through the week to a peak on Saturday; however the Sunday figures are much lower.

The newspaper company want to see the underlying trend and so they look at the 7-point running medians. The median for Week 1 is 5.6 and this is shown in the table below at the middle point of that week, i.e Thursday. The medians for each subsequent 7-day period are similarly shown on the day at the middle of the period.

	Mon	Tues	Wed	Thurs	Fri	Sat	Sun
Week 1	–	–	–	5.6	5.8	6.1	6.2
Week 2	6.2	6.2	6.2	6.2	6.6	6.8	6.9
Week 3	6.9	7.0	7.0	7.0	–	–	–

The upward sales trend can be seen in Fig. 108.

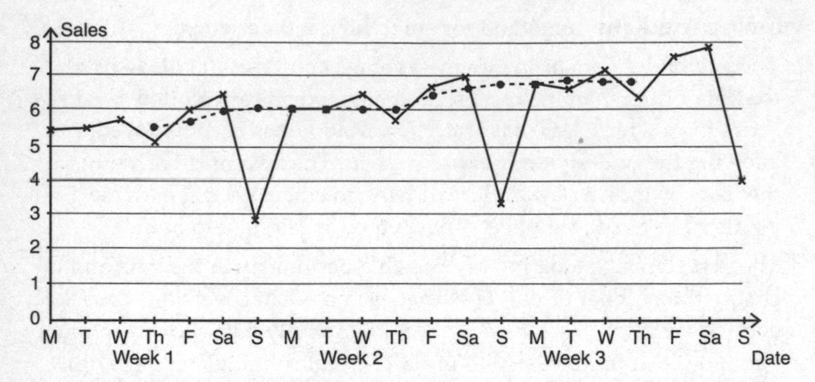

Fig. 108. **Running medians.** Actual Data ——— Running Median •-------•

See MOVING AVERAGE.

S

sample a set of individuals or items selected from a parent population to provide information about its distribution. This information may be the general shape and properties of the distribution or estimates of its parameters; alternatively the sample data may be used for testing hypothesis about aspects of the population.

A sample is taken with a target population in mind. The more representative it is of the target population, the more useful it is, and there are various sampling methods which are designed with this intention. See CLUSTER SAMPLING, OPPORTUNITY SAMPLING, QUOTA SAMPLING, SIMPLE RANDOM SAMPLING, STRATIFIED SAMPLING, SYSTEMATIC SAMPLING. See also EXPERIMENTAL DESIGN and FRAME.

The distribution of the means of samples of a particular size is called the sampling distribution of the means; similarly the distribution of the variances of the samples is the sampling distribution of the variances; and so on. These sampling distributions form the basis for hypothesis tests on their parameters. See SAMPLING DISTRIBUTION, SIMPLE RANDOM SAMPLE.

The following population parameters are commonly estimated from samples.

From a single sample $(x_1, x_2, \ldots x_n)$

$$\text{Estimated population mean} = \frac{\sum x_i}{n} = \overline{x}$$

$$\text{Estimated population variance} = \frac{\sum (x_i - \overline{x})^2}{n-1} = s^2$$

From two samples, sizes n_1 and n_2, from the same population

$$\text{Estimated population mean} = \frac{n_1\overline{x}_1 + n_2\overline{x}_2}{n_1 + n_2}$$

$$\text{Estimated population variance} = \frac{(n_1 - 1)s_1^2 + (n_2 - 1)s_2^2}{(n_1 + n_2 - 2)}$$

For k samples from the same population

$$\text{Estimated population mean} = \frac{n_1\bar{x}_1 + n_2\bar{x}_2 + \ldots + n_k\bar{x}_k}{n_1 + n_2 + \ldots n_k}$$

$$\text{Estimated population variance} = \frac{(n_1 - 1)s_1^2 + (n_2 - 1)s_2^2 + \ldots + (n_k - 1)s_k^2}{(n_1 + n_2 + \ldots + n_k - k)}$$

The distinction is sometimes made between sampling from finite and infinite populations. The formulae above refer to supposedly infinite populations although in reality this means large, or very large, since all populations are actually finite in size. For the results for smaller finite populations see FINITE POPULATION CORRECTION FACTOR.

Another distinction that is occasionally made is between sampling with or without replacement. (Sampling with replacement would make the population effectively infinite.) However, in practice sampling is almost always carried out without replacement (because it usually makes no sense to ask the same individual for information twice).

sample point an element in a sample space, a possible outcome of an experiment.

sample size the number of individuals or items in a sample. Samples are often called small if their size is under about 30.

sample space the set of possible outcomes of an experiment; the values that can be taken by a random variable. See also OUTCOME.

sampling distribution the distribution of a statistic obtained from a sample of a particular size from a given population. The sampling distribution of the means is described by the central limit theorem.

sampling error the difference between the true value of a population parameter and that estimated from a sample. This error is due to the fact that the value has been calculated from a particular sample.

Example: a large shoal of fish has mean mass 0.92 kg and standard deviation 0.25 kg. A fisherman catches five fish from the shoal with masses 0.81, 0.63, 1.04, 1.11, and 0.91 kg. The mean mass of his fish is 0.90 kg and he uses this figure as an estimate for the mean mass of the shoal. The sampling error for the mean is:

$$0.92 - 0.90 = 0.02 \text{ kg}$$

He also estimates the standard deviation of the shoal to be 0.19 kg using

$$\sqrt{\frac{\sum(x_i - x)^2}{n-1}}$$

as an estimator. The sampling error for the standard deviation is thus

$$0.25 - 0.19 = 0.06\text{kg}$$

sampling fraction at its simplest,

$$\text{sampling fraction} = \frac{\text{sample size}}{\text{population size}}.$$

Thus sampling fraction is the proportion of the population which is sampled. This definition cannot, however, be applied to all sampling schemes. In stratified sampling, for example, each stratum has its own sampling fraction.

sampling line the line on a lattice diagram illustrating good and defective samples. See LATTICE DIAGRAM.

scatter diagram a graph illustrating a sample from a bivariate distribution. The points are plotted but not joined.

A scatter diagram is usually drawn before working out a correlation coefficient; it gives a good initial indication of whether there is any value in doing this calculation. If the points lie roughly on a straight line, it is reasonable to go ahead, but if they are scattered all over the graph paper, the calculation is likely to be a waste of time.

A scatter diagram will also show up cases where calculation of a correlation coefficient is not appropriate, for example: a non-linear relationship (see Fig. 109(a)); a distribution where the calculation would be distorted by outliers (see Fig. 109(b)); a distribution in which the data appear to fall into two or more distinct areas (see Fig. 109 (c)).

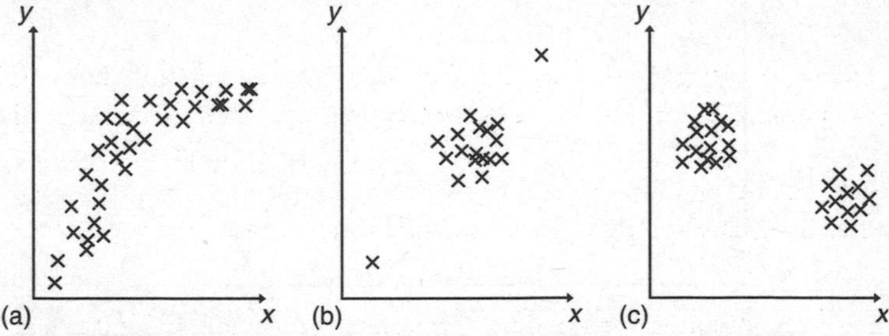

Fig. 109. **Scatter diagram.** Some cases where calculation of a correlation coefficient is not appropriate.

seasonal component the component of variation in a time series which is dependent on the time of year. Symbol: S. The costs of various types of fruit and vegetable, unemployment figures, and mean daily rainfall, all show marked seasonal variation. Compare IRREGULAR VARIATION. See TIME SERIES.

semi-averages, method of a method for estimating trend. The data are in the form of a time series; the figures are usually divided chronologically into two groups of equal size, and the means of each taken. These means are then used to estimate the trend.

Example: the passenger miles travelled on railways within the UK between 1993 and 2002, in billions:

Year	1993	1994	1995	1996	1997	1998	1999	2000	2001	2002
Journeys	37	35	37	39	42	44	46	47	47	48

The mean for the first 5-year period (1993-1997) = 38
The mean for the second 5-year period (1998-2002) = 46.4
The means of the two sets of years are 1995 and 2000.

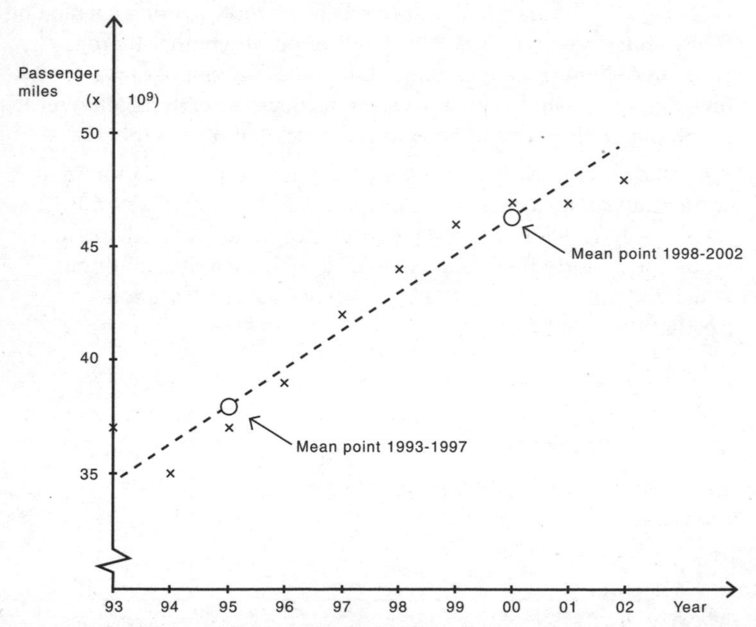

Fig. 110. **Semi-averages, method of.** Graph illustrating the trend of railway passenger miles ($\times 10^9$), 1993-2002.

So the trend is estimated as:

$$\frac{46.4 - 38}{2000 - 1995} = \frac{8.4}{5} = 1.68$$

this represents an increase of 1.68 billion passenger miles per year, as shown in Fig. 110.

See also LINE OF BEST FIT.

semi-interquartile range a measure of spread, or dispersion, given by:

$$\frac{\text{upper quartile-lower quartile}}{2}$$

See also INTERQUARTILE RANGE.

semi-logarithmic graph see GRAPH.

sequential analysis, a technique for hypothesis testing in which observations are taken one batch at a time (the batch may be a single item) and at each stage one of three possible decisions is made:

- to accept the null hypothesis;
- to reject the null hypothesis;
- to decide that there is not enough yet enough evidence, and therefore to continue sampling.

If such a procedure is designed well, it can give a test having better Type 1 and Type 2 error probabilities than a conventional test with a fixed sample size. Alternatively it can result in a test having a smaller expected sample size for the same Type 1 and Type 2 error probabilities than a fixed sample size test.

Such procedures are widely used in some aspects of quality control and other industrial applications of statistics; a double sampling scheme for quality control is a simple example of a sequential procedure. Such procedures are also sometimes used in other areas of statistics. See LATTICE DIAGRAM, TYPE 1 AND TYPE 2 ERRORS.

sequential probability ratio test a procedure in sequential that is used to determine whether to accept the null hypothesis, reject it or continue testing.

set a collection of items that usually have at least one common property or characteristic. The members of a set are called *elements*. By convention, the brackets used to indicate a set are written { }.

A diagram illustrating the relationship between sets is called a *Venn diagram*, also called an Euler diagram.

Set notation, and the associated language, is often used in statistics, for a number of reasons:

- It provides a concise way of describing particular groups and the relationships between them;
- the description of events is closely related to that of sets;
- the number of elements in a set may be needed for a probability calculation based on relative frequency.

Example: This entry is based on the theme of a scientist investigating the rabbit population of Fair Isle. Fair Isle is a remote and isolated island off the north coast of Scotland; many of its rabbits have unusual colours.

The set from which members of a particular set may be drawn is called the *universal set*, denoted by $\mathcal{E}$. In this entry $\mathcal{E}$ = {Rabbits living on Fair Isle}.

Other sets used are:

M = {Males}, F = {Females}, B = {Black rabbits}, R = {Orange rabbits},

L = {Rabbits weighing over 3 kg}

The symbol $\in$ means 'is an element of' or 'is a member of'. Thus Rabbit A32 $\in F$ reads as "Rabbit A32 is a member of the set of females" and, since $\mathcal{E}$ = {Rabbits living on Fair Isle}, it is understood that the females in question are rabbits living on Fair Isle. So A32 is a female rabbit living on Fair Isle.

The *union* of two sets, denoted by the symbol $\cup$, is the set of all elements in one or other or both of the sets. So the set $F \cup B$ is those rabbits which are female, or black, or are both female and black. This is shown by the shaded region in the Venn diagram in Fig. 111.

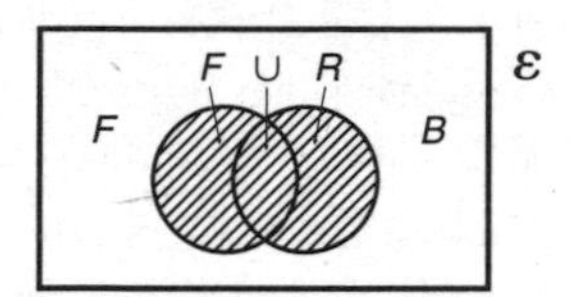

Fig. 111. **Set.** Union of two sets.

The *intersection* of two sets, denoted by the symbol $\cap$, is the set of all elements in both of the sets. So the set $F \cap R$ consists of rabbits which are both female and orange. This is shown by the shaded region in Fig. 112.

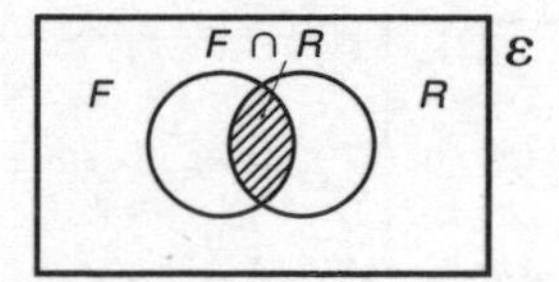

Fig. 112. **Set.** Intersection of two sets.

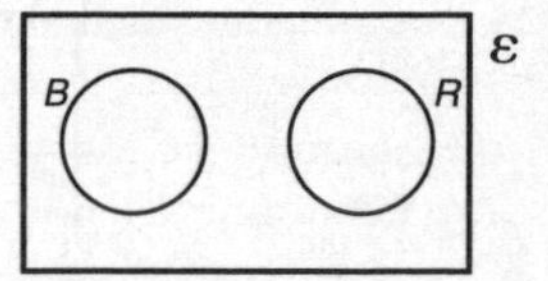

Fig. 113. **Set.** Disjoint sets.

A set with no members is called the empty set and denoted by the symbol $\varnothing$. Fig. 113 shows the relationship between the sets *B* and *R*. The sets do not intersect and so the set $B \cap R$ is the empty set. In set notation this is written $B \cap R = \varnothing$; in English it reads "There are no rabbits which are both black and orange". Sets which do not intersect are called *disjoint*.

By contrast Fig. 112 shows that the sets *F* and *R* do intersect. This is written $F \cap R \neq \varnothing$ and reads as "There are some females which are orange".

If every element of one set also belongs to another set, then the smaller set is described as a *subset* of the larger. The symbol $\subset$ means "is a subset of". Thus Fig. 114 illustrates that $L \subset M$ and reads "All the rabbits weighing over 3 kg are male".

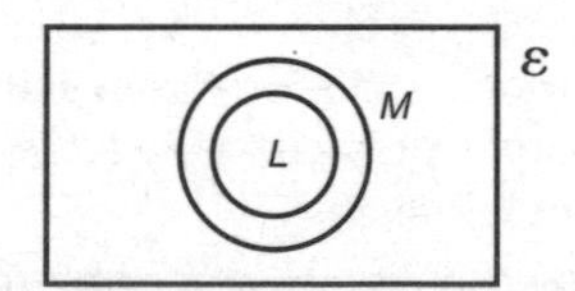

Fig. 114. **Set.** *L* is a subset of set *M*.

The *complement* of a set is the set of all elements in the universal set but not in that set. So the complement of *B*, denoted by B', is the set of all the rabbits that are not black. See Fig. 115.

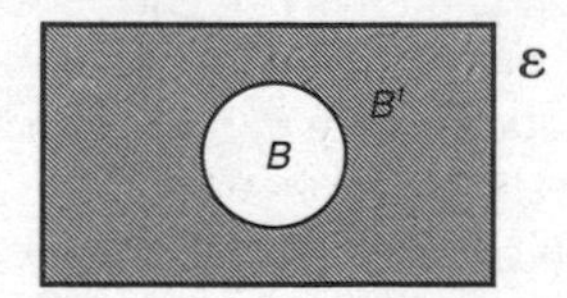

Fig. 115. **Set.** Set *B* and its complement B^{1}.

Two or more disjoint sets whose union is the universal set are described as *complementary*. The sets *F* and *M* are complementary because $F \cap M = \varnothing$ and $F \cup M = \mathcal{E}$. No rabbit on Fair Isle is both male and female, and all rabbits are either male or female. See Fig 116.

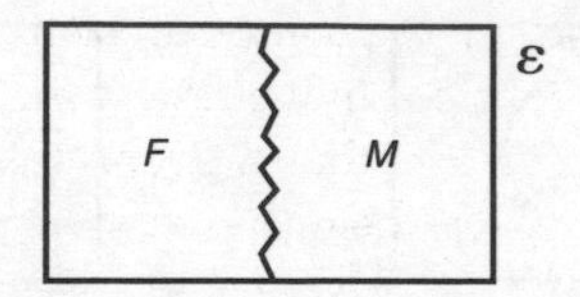

Fig. 116. **Set.** Complementary sets.

The letter n is used to denote the number of elements in a set. Thus the number of black rabbits is $n(B)$ numbers and the probability that a random selected rabbit is black is given by

$$\frac{n(B)}{n(\mathcal{E})}$$

A useful relationship is that for any two sets, X and Y,

$$n(X) + n(Y) = n(X \cup Y) + n(X \cap Y).$$

Sheppard's correction see GROUPED DATA.

Shewhart means chart see CUSUM CHARTS.

sigma ($\sum$) notation see SUMMATION NOTATION.

significance level the probability of rejecting a true null hypothesis in a statistical test.

It is not possible to draw conclusions from statistical data in the way that one can, for example, prove that the angle sum of a triangle is 180°, i.e. with absolute certainty.

Suppose a coin is tossed 100 times and comes down heads every time. There is clearly strong reason to suspect the coin of being biased, but there is still the remote possibility that it is unbiased and just happened to fall that way. What can be said is that the probability of a fair coin landing the same way 100 times out of 100 is $2\times(\frac{1}{2})^{100}$, which is 1.6×10^{-30}. So if, on this evidence, the null hypothesis 'The coin is unbiased' is rejected in favour of the alternative hypothesis 'The coin is biased', there is still this very small probability that the conclusion is incorrect.

In that case, the probability of error is clearly so small that we would have no hesitation in declaring the coin biased, even though the conclusion could theoretically be wrong. If, however, the coin had been tossed 8 times and came down the same way each time, we might think rather more carefully before declaring it biased. The probability of a fair coin landing the same way 8 times out of 8 is 1/128; it is unlikely, but not entirely negligible.

The significance level of a test is the probability at which we decide to reject the null hypothesis. In the case of the coin, this would mean declaring it to be biased. It is thus the probability of coming to a wrong conclusion, making the mistake of rejecting a true null hypothesis that we are prepared to accept. If the probability of a result at least as extreme as that found is less than or equal to the significance level, the null hypothesis is rejected.

When a null hypothesis which is in fact true is rejected, a Type 1 error occurs. When a false null hypothesis is accepted, a Type 2 error occurs. For a given sample size, the probabilities of each of these types of error are related. The lower the significance level of the test, the smaller is the probability of a Type 1 error, but the greater that of a Type 2 error. The significance level chosen for any particular test depends on the consequences of each type of error.

Example: a sample of the rivets on the wing of an aircraft is tested for possible weakening.

Null hypothesis, H_0: the rivets are of the correct strength.
Alternative hypothesis, H_1: the rivets have become weaker.

Type 1 error: the rivets are all right, but declared to be weak.
Type 2 error: the rivets are weak, but declared to be all right.

The consequence of a Type 1 error would be that expensive maintenance work is carried out unnecessarily.

The consequence of a Type 2 error would be that the aircraft's safety is in danger.

In this case a Type 1 error is clearly more desirable (or less undesirable) than a Type 2 error, and so the significance level would be set at quite a high figure.

Statistical tests are usually carried out at predetermined significance levels. Typical statements of results might be: 'There is no reason, at the 10% significance level, to reject the null hypothesis that there is no difference in height between the adult men in Accra and Lagos'. 'The claim that the machine is producing nails shorter than the specified length is accepted at the 5% significance level: the null hypothesis that their length is correct is rejected'.

In many tests the probability need not actually be worked out. The test data are used to work out a statistic, like X^2 or t, which is compared to a critical value found in tables, for the required significance level and degrees of freedom.

significant figures **1.** the figures of a number that express its magnitude to a specified degree of accuracy.

2. the number of such figures.

Example: 3.1415927 written to 4 significant figures is 3.142. The digits, 3, 1, 4 and 2, are the four significant figures.

sign test **1.** a non-parametric test of the null hypothesis that two matched samples are drawn from the same population. It actually tests for a median difference of zero.

Example: a sample of 10 patients in a hospital were given a new sleeping pill. The number of hours they slept that night were recorded and compared to the figures for the previous night, when they had taken the old pill. Is there reason, at the 5% significance level, to think the new pill different from the old one in effect?

H_0: There is no difference in the effects of the two pills.
H_1: There is a difference in the effects of the two pills.

2-tail test

5% significance level

	Old pill	*New pill*	*Sign of difference*
A	8	$8^1/_4$	+
B	6	$4^1/_2$	–
C	7	7	0
D	5	9	+
E	7	$7^1/_2$	+
F	$6^1/_2$	$6^1/_4$	-
G	8	8	0
H	7	$7^1/_2$	+
I	4	5	+
J	6	$8^1/_4$	+

In the right hand column those who slept longer have been given +, shorter –, and those who slept for the same hours, 0.

The two who slept the same length of time, C and G, are discounted from the sample. Thus there remain 8 members, 6+ and 2–.

The null hypothesis can now be restated as P(positive result, +) = P(negative result, –) = $^1/_2$.

It is now possible to apply the binomial distribution to this situation (see also PASCAL'S TRIANGLE).

Outcome	8+, 0–	7+, 1–	6+, 2–	5+, 3–	4+, 4–
Probability	$\frac{1}{2^8}$	$\frac{8}{2^8}$	$\frac{28}{2^8}$	$\frac{56}{2^8}$	$\frac{70}{2^8}$

Outcome	3+, 5–	2+, 6–	1+, 7–	0+, 8–
Probability	$\frac{56}{2^8}$	$\frac{28}{2^8}$	$\frac{8}{2^8}$	$\frac{1}{2^8}$

The actual result obtained was 6+, 2–. The results as far or further from the central value of 4+, 4– are 8+, 0–, or 7+, 1–, or 6+, 2–, or 2+, 6–, or 1+, 7– or 0+, 8–.

$$\frac{1}{2^8}+\frac{8}{2^8}+\frac{28}{2^8}+\frac{28}{2^8}+\frac{8}{2^8}+\frac{1}{2^8}=\frac{74}{256}=0.29$$

The total of their probabilities is:

Since 0.29 > 0.05, the 5% significance level, there is no reason to reject the null hypothesis. There is insufficient evidence to support the alternative hypothesis that there is a difference in the effects of the two pills.

If cumulative probability tables for the binomial distribution (for $p = {}^1/_2$) are available, the calculation can largely be avoided.

The sign test is very crude. In the example, no account was taken of how much longer, or shorter, the various patients slept. A slept only $^1/_4$ hour more, D 4 hours, but both are given the same rating, just +. It is thus not particularly well-suited for this sort of situation, involving quantitative data. A Wilcoxon matched-pairs signed-rank test or a paired sample *t*-test would have been more appropriate. Had the patients, on the other hand, been asked a qualitative question like, 'Was your sleep more relaxed?', the sign test would have been quite suitable, counting + for 'Yes', - for 'No', and 0 for 'No change'.

2. a test for a single median.

Example: the label on a match box (of a particular brand) carries the statement "Contents 50 matches". It is suspected that the median number of matches is actually less than 50. A random sample of 10 match boxes is taken and the matches in each of them are counted:

51, 48, 49, 50, 49, 50, 50, 48, 49, 47.

The sign test may then be used as a hypothesis test.

H_0: The population median number of matches is 50.

H_1: The median is less than 50.

1-tail test

10% significance level (say).

The test then proceeds as for the paired sample case. The data are compared to the null hypothesis value of 50, giving:

+ , - , - , 0 , - , 0 , 0 , - , - , -

In this case there are 6 – results and 1 + result.

The probability of this result, or the more extreme one of 6 – and no + is given by

$$7 \times \left(\tfrac{1}{2}\right)^6 \times \left(\tfrac{1}{2}\right) + \left(\tfrac{1}{2}\right)^7 = \tfrac{8}{128} = 0.0625$$

and since 0.0625 < 0.10 (i.e. 10%), the null hypothesis is rejected. The evidence supports the claim that the median number of matches is less than 50, at the 10% significance level.

simple random sampling sampling in which every possible sample of a given size has the same probability of being chosen. It can be proved that this is equivalent to each member of the parent population having an equal probability of being selected provided that each selection is independent of the others, the population is finite and the sampling is without replacement.

By contrast, in *random sampling* (as usually understood) it is only necessary that every member of the population has an equal probability of selection but not that they are independent. Thus in systematic sampling, with a random start, and stratified sampling with sub-sample sizes proportional to the strata sizes, each element at the outset is equally likely to be selected. However in both of these cases there are many samples of the parent population which could not be obtained because the items are not selected independently of one another.

simulation the use of a mathematical model to reproduce the conditions of a situation or process, and to carry out experiments on it. Simulation may be carried out for reasons of economy, because it is difficult (or impossible) to conduct an experiment or because a theoretical analysis is too difficult.

Simulation of statistical process or experiments is often done by computer using random numbers.

single sampling scheme in quality control, a sampling scheme where only one sample per batch is inspected. If the number of defectives found is not more than the acceptance number, the batch is accepted. Otherwise it is rejected, or subjected to a 100% inspection.

skew (of a distribution) not having equal probabilities above and below the mean. A skew distribution can be described as having *skewness*.

Population skewness is measured by the *coefficient of skewness*:

$$E\left(\frac{X-\mu}{\sigma}\right)^3.$$

See EXPECTATION.

In a skew distribution, the median and the mean are not coincident. If the median is less than the mean, the distribution has positive *skewness*, and vice versa, as in Fig. 117.

If there is only one mode, the median is found empirically to lie between the mode and the mean, the relationship given approximately by:

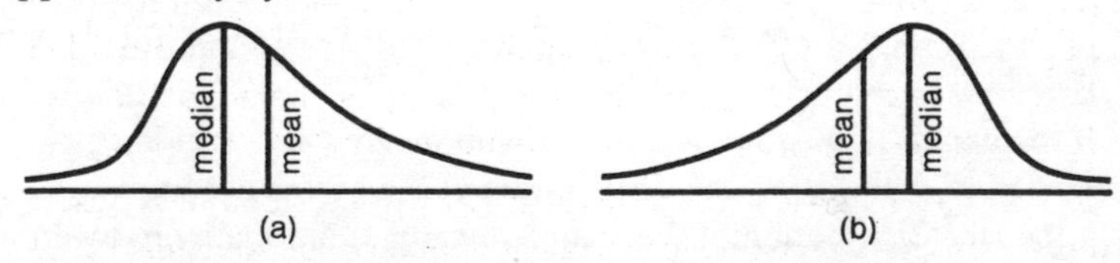

Fig. 117. **Skew.** (a) Positively skew distribution, where the median<mean. (b) Negatively skew distribution, where the median>mean.

$$mode = median - 3 \times (mean - median)$$

Thus, for a unimodal distribution, it is also true that:

$$\text{Mode} < \text{mean} \Rightarrow \text{Positive skewness}$$
$$\text{Mode} > \text{mean} \Rightarrow \text{Negative skewness}$$

This underlies *Pearson's measure of skewness* which is applied to sample data and is given by:

$$\frac{\text{mean} - \text{mode}}{\text{standard deviation}} \quad \text{or} \quad \frac{3(\text{mean} - \text{median})}{\text{standard deviation}}$$

Another measure of skewness applied to sample data is the *quartile coefficient of skewness*, given by:

$$\frac{\text{upper quartile} + \text{lower quartile} - 2 \times \text{median}}{\text{upper quartile} - \text{lower quartile}}$$

Distributions with positive and negative skewness give distinctive non-straight line graphs when drawn on Normal probability graph paper (Fig. 118).

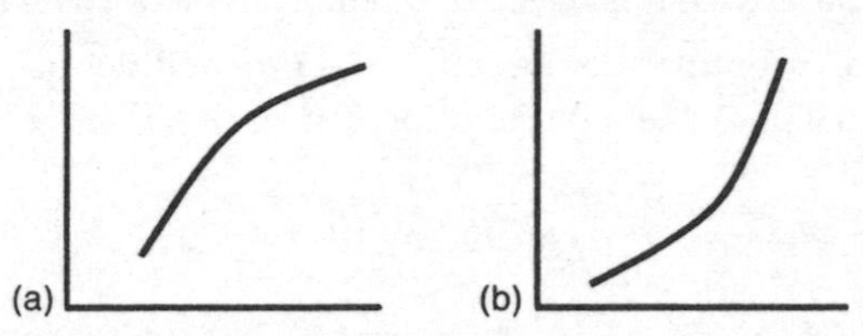

Fig. 118. **Skew.** Lines when distributions are plotted on Normal probability graph paper. (a) Positively-skew distribution. (b) Negatively-skew distribution.

slit chart see PIE CHART.

small sample a sample from which, because of its small size, conclusions have to be treated with particular caution. Hypothesis testing may involve the use of special tests.

A sample is often considered small if it is less than about 30 in size. This does, however, depend on the reason why the sample is being taken; what would be a large sample under one set of circumstances may be a small one under others.

If the parent population is Normal with mean μ and standard deviation σ, the sampling distribution of the means of samples of size n (i.e., the distribution of the sample means) is Normal with mean μ, and standard deviation $\frac{\sigma}{\sqrt{n}}$, where n is the sample size, whether large or small.

It is, however, often the case that the parent standard deviation is unknown and can only be estimated from the sample. If this estimate, s, is used instead of σ (the true parent standard deviation), the sampling distribution of the sample means is the t distribution with n-1 degrees of freedom rather than the Normal distribution (assuming the underlying population is itself Normal).

If the sample size is large, the t distribution is almost the same as the Normal distribution, but for a small sample it is markedly different.

Spearman's rank correlation coefficient see CORRELATION COEFFICIENT.

special cause see ASSIGNABLE CAUSE.

split-half method see RELIABILITY COEFFICIENT.

spread see DISPERSION.

square law graph see GRAPH.

standard deviation a measure of spread or dispersion. Notation: σ (population standard deviation), s (estimated population standard deviation). Standard deviation is the square root of variance.

The population standard deviation of a random variable X is defined as:

$$\sigma = \sqrt{\left(\mathrm{E}(X^2) - [\mathrm{E}(X)]^2\right)}.$$

See EXPECTATION.

- If X is a continuous variable with probability density function f(x), this is

$$\sigma = \sqrt{\left(\int x^2 \mathrm{f}(x)\,\mathrm{d}x - [\mathrm{E}(X)]^2\right)}$$

- The equivalent formula for a discrete variable, X, with probability distribution p(x_i), where $1 \le i \le n$ and n is a positive integer, is

$$\sigma = \sqrt{\sum_i \left(x_i^2 \mathrm{p}(x_i) - [\mathrm{E}(X)]^2\right)}$$

The calculation of sample standard deviation, and indeed its very meaning, is complicated by the use of two different formulae giving two different answers. Both refer to a set of data $x_1, x_2, \ldots x_n$, with mean $\overline{x}$.

Divisor n-1

$$\text{Standard deviation, } s = \sqrt{\frac{1}{n-1}\sum (x_i - \overline{x})^2}$$

Divisor n

$$\text{Standard deviation } \sqrt{\frac{1}{n}\sum (x_i - \overline{x})^2}$$

In this case there is no recognised notation for standard deviation.

In advanced statistics, the (n-1) form is almost universally used. Unless otherwise stated, this is the form used in this book.

Both forms give the square root of an average of the square of the deviations of the data values from their mean, $(x_1 - \bar{x})^2$. In the one case the average is found by dividing by the number of independent deviations, n-1; one has been lost in determining the mean. In the other case the average is found simply by dividing by the number of deviations, n.

The square of standard deviation is the variance and so this too has two forms, one with divisor n-1 and the other with divisor n.

The form of sample variance calculated with divisor n-1,

$$s^2 = \frac{\sum (x_i - \bar{x})^2}{n-1}$$

is an unbiased estimator for the parent population variance making it an important statistic in its own right.

As well as the formulae already given there are equivalent formulae that are used for calculating standard deviation and variance, given in terms of the sum of squares, S_{xx}, where

$$S_{xx} = \sum_{i=1}^{n} (x_i - \bar{x})^2 ;$$

an equivalent form of S_{xx} is

$$S_{xx} = \sum_{i=1}^{n} x_i^2 - \frac{1}{n}\left(\sum_{i=1}^{n} x_i\right)^2 .$$

Using S_{xx}, standard deviation is given by:

Divisor n-1

$$s = \sqrt{\frac{S_{xx}}{n-1}}$$

Divisor n

$$\sqrt{\frac{S_{xx}}{n}}$$

For data sorted into k groups,

Value	x_1	x_2	...	x_k
Frequency	f_1	f_2	...	f_k

with $f_1 + f_2 + \ldots + f_k = n$

the equivalent forms of S_{xx} are written:

$$S_{xx} = \sum_{i=1}^{k} (x_i - \bar{x})^2 f_i \quad \text{and} \quad S_{xx} = \sum_{i=1}^{k} x_i^2 f_i - \frac{1}{n}\sum_{i=1}^{k} (x_i f_i)^2 .$$

Example: For the figures, 12, 9, 9, 9, 8, 6, 6, 5, 4, 2, the calculation of standard deviation using the first of the formula for grouped data is as follows:

x_i	f_i	x_if_i	$x_i - \bar{x}$	$(x_i - \bar{x})^2$	$(x_i - \bar{x})^2 f_i$
12	1	12	5	25	25
9	3	27	2	4	12
8	1	8	1	1	1
6	2	12	-1	1	2
5	1	5	-2	4	4
4	1	4	-3	9	9
2	1	2	-5	25	25
$k = 7$	$n = 10$	70			$S_{xx} = 78$

Standard deviation (divisor n-1) $s = \sqrt{\frac{78}{9}} = 2.94$ (3 s.f.)

(divisor n) $= \sqrt{\frac{78}{10}} = 2.79$ (3 s.f.)

See SUM OF SQUARES.

standard error **1.** the standard deviation of the means of samples of a given size, drawn from a particular parent population; the standard deviation of the sampling distribution of the means.

For a large population with standard deviation σ, or for sampling with replacement, the standard error for samples of size n is given by:

$$\frac{\sigma}{\sqrt{n}}.$$

For a population of size N, standard error is given by:

$\frac{\sigma}{\sqrt{n}}\sqrt{1-\frac{n}{N}}$ which can be written as $\frac{\sigma}{\sqrt{n}}\sqrt{1-f}$ where f is the sampling fraction. (See SAMPLING FRACTION).

Standard error is a measure of a reasonable difference between a sample mean and the parent mean, and is used in tests of whether a particular sample could have been drawn from a given parent population. It is also used in working out confidence limits and confidence intervals.

There is some confusion in the nomenclature and notation used for standard error of the mean. Some people use it to mean either $\frac{\sigma}{\sqrt{n}}$ or,

in cases where σ is unknown and estimated by s, $\frac{s}{\sqrt{n}}$. Others use standard error for $\frac{\sigma}{\sqrt{n}}$ and estimated standard error for $\frac{s}{\sqrt{n}}$.
A third group reserve the term standard error for $\frac{s}{\sqrt{n}}$ and refer to $\frac{\sigma}{\sqrt{n}}$ as the standard deviation of $\bar{X}$.

2. the standard deviation, or estimated standard deviation, of sample statistics other than the mean.

standard form a way of writing numbers which is particularly convienient if they are very large or very small. A number in standard form is of the type, $a \times 10^n$, where a lies between 1 and 10, and n is an integer. The mass of an electron is actually

0.000 000 000 000 000 000 000 000 000 000 911kg

This is more conveniently expressed in standard form as 9.11×10^{-31}kg. (The decimal point has been moved 31 places to the right).

standardisation transformation of the variable in a distribution, so that it has mean 0 and standard deviation 1. Standardisation is carried out using the transformation

$$z = \frac{(x - \mu)}{\sigma}$$

where μ is the mean, σ the standard deviation, x the original value, and z the new value.

Example: a machine makes bolts of mean length 10 cm, with standard deviation 0.05 cm. A particular bolt has length 10.12 cm. The standardised form of this length is:

$$z = \frac{(10.12 - 10.00)}{0.05} = 2.4$$

Standardisation allows general tables for certain distributions to be constructed and used, like those for the Normal distribution and the *t* distribution. See Tables 1, 2 & 3.

statistic any function of sample data, containing no unknown parameters, such as mean, median or standard deviation.

The term statistic is used for a sample. By contrast the term parameter is used for a parent population. By convention, parameters are often assigned Greek letters (like μ and σ), statistics Roman letters (e.g. m and s).

stem-and-leaf plot or **stemplot** a diagram used to illustrate data. Stem-and-leaf plots are used to bring out the major features of the distribution of the data in a convenient and easily drawn form; they are used in exploratory data analysis.

Example: the stem-and-leaf plot showing the birth rates in 18 European countries, per thousand population (2003).

Austria	9.4	Denmark	11.5	Finland	10.5	France	12.5
Germany	8.6	Greece	9.8	Hungary	9.3	Ireland	14.6
Italy	9.2	Luxembourg	11.9	Netherlands	11.3	Norway	12.2
Poland	10.5	Portugal	11.4	Romania	10.8	Sweden	9.7
Switzerland	9.6	UK	11.0			$n = 18$	

Stem	*Leaves*
8	6
9	2 3 4 6 7 8
10	5 5 8
11	0 3 4 5 9
12	2 5
13	
14	6

12|5 means 12.5
Legend

The figures to the left of the line are called the stem, those to the right are leaves. For example, against the stem figure of 12 are entered the figures for Norway (12.2) and France (12.5) in the form of leaf figures 2 and 5. To make the diagram more compact, outliers are often listed separately, low values at the low end and high values at the high end.

A legend should always accompany a stem-and-leaf plot.

In a back-to-back stem-and-leaf plot, two sets of data share the same stem. The leaves of one appear to the left and of the other to the right. This allows the two sets of data to be compared. One set might, for example, refer to females and the other to males.

stemplot see STEM-AND-LEAF PLOT.

Stirling's approximation see FACTORIAL.

stochastic (of a variable) exhibiting random behaviour, which cannot be explained fully be a purely deterministic model.

stochastic process a random process, usually a variable measured at a set of points in time or space, which can be in one of a number of states. For example, the length of a queue measured at different times is a stochastic process. In the special case when the next state depends only on the present state, and not on the past history of the process, a stochastic process is called a Markov chain.

The probability that a system in a particular state at one time will be in some other state at a later time is given by the *transition probability*.

stratified sampling a method of sampling which makes allowance for known differences within the underlying population.

Example: an opinion pollster is carrying out a survey of voting intentions in one area before a general election. She knows that voters can be divided into three strata: rural (30%), suburban (50%) and urban (20%) populations. So she ensures that 30% of those she samples are rural voters, 50% suburban and 20% urban. This is a stratified sample. It would be usual to use simple random sampling in making the selection within each stratum. In this example, the proportions from each of the three strata in the sample are the same as those in the strata in the population. This is described as *proportional allocation*; it is not an essential feature of stratified sampling.

Stratified sampling usually reduces the variability of the estimates produced from the sample compared to simple random sampling, and so is preferable in situations where it can be used. Another advantage is that it provides information about the separate strata.

Student's t test see *t* TEST.

subset see SET.

summation notation the use of the symbol $\sum$ to 'mean the sum of ...'. For example, $x_1 + x_2 + ... + x_n$ in summation notation is written

$$\sum_{i=1}^{n} x_i$$

This is often written more loosely as,

$\sum_i x_i$, $\sum x_i$, or just $\sum x$ meaning the sum of all the values of x.

Example: If x has values 2, 3, 8, 10 and 11, then

$$\sum x = 2+3+8+10+11 = 34$$
$$\sum x^2 = 4+9+64+100+121 = 298$$

sum of squares and sum of products the term sum of squares usually means the sum of the squares of the deviations from the mean. Notation S_{xx}. This can be calculated as:

$$S_{xx} = \sum_{i=1}^{n}(x_i - \bar{x})^2 \text{ or } S_{xx} = \sum_{i=1}^{n} x_i^2 - \frac{1}{n}\left(\sum_{i=1}^{n} x_i\right)^2 = \sum_{i=1}^{n} x_i^2 - n\bar{x}^2$$

or, for grouped data, $x_1, x_2, \ldots x_k$ with frequencies $f_1, f_2, \ldots f_k$, (where $n = \sum_{i=1}^{k} f_i$):

$$S_{xx} = \sum_{i=1}^{k}(x_i - \bar{x})^2 f_i \text{ or } \sum_{i=1}^{k} x_i^2 f_i - \frac{1}{n}\left(\sum_{i=1}^{k} x_i f_i\right)^2 = \sum_{i=1}^{k} x_i^2 f_i - n\bar{x}^2$$

Example: calculate the sum of the squares for the numbers 4, 4, 5, 8, 9.

x	f	xf	$x-\bar{x}$	$(x-\bar{x})^2 f$
4	2	8	-2	8
5	1	5	-1	1
8	1	8	2	4
9	1	9	3	9
	$n=5$	$\sum xf = 30$		$S_{xx} = 22$

Mean $\bar{x} = \dfrac{\sum xf}{\sum f} = \dfrac{30}{5} = 6$

Sum of squares $S_{xx} = 22$

The sum of squares is sometimes called the *corrected sum of squares*. The quantity $\sum x_i^2$ is sometimes called the *uncorrected sum of squares* or the *crude sum of squares*.

The sum of products, S_{xy}, is given by $S_{xy} = \sum_{i=1}^{n}(x_i - \bar{x})(y_i - \bar{y})$.

See CORRELATION, REGRESSION LINE, STANDARD DEVIATION, VARIANCE.

sum of two or more random variables see DIFFERENCE AND SUM OF TWO OR MORE RANDOM VARIABLES.

superior a term used in clinical trials. Confidence intervals for the effects of three treatments, measured on a suitable scale, are illustrated in Fig. 119.

Fig. 119. **Superior.** Treatment *A* is superior to treatment *C* but non-superior to treatment *B*.

Treatment *A* is described as *superior* to treatment *C* because their confidence intervals do not overlap. Correspondingly treatment *C* is described as *inferior* to treatment *A*.

By contrast treatment *A* is *non-superior* to treatment *B*; their confidence intervals overlap and so it is possible (at the stated level of confidence) that treatment *B* is at least as good as treatment *A*. In the same way treatment *B* is *non-inferior* to treatment *A*.

The term *equivalent* is used to describe an outcome of a trial comparing a proposed new treatment with the standard treatment: the new treatment is equivalent if its confidence interval is contained entirely within the confidence interval of the standard treatment.

survey an investigation of one or more variables of a population, which may be quantitative or qualitative. Surveys are often, but by no means invariably, carried out on human beings. A survey may involve 100% sampling, in which case it is called a census, for example, a survey of the opinions of all the householders of a village about a proposed by-pass. When this is not the case, the term does, nonetheless, imply adequate sampling. It is common to use a questionnaire for collecting information in a survey; this is often devised in the light of evidence from a pilot survey.

sweeping by means a technique for fitting a linear model in a situation where the response variable is a function of two categorical variables. This technique is very like median polish but the column and row means are reduced by the means rather than the medians. (See MEDIAN POLISH.)

Sweeping by means underlies the standard two-way analysis of variance, and is used in the programming of some computer packages.

systematic error error due to the method of collecting or processing data. This may be due to bias in the EXPERIMENTAL DESIGN, or to the ESTIMATOR being used.

systematic sampling sampling in which each member of the sample is chosen by ordering the frame in some way, and then selecting individuals at regular intervals. Thus a 1% sample of telephone subscribers could be taken by selecting the 21st, 121st, 221st, 321st (and so on) of the names listed in the telephone directory.

A systematic sample is often given a random start. In the case of a 1% sample, a random number would be chosen between 1 and 100, like the number 21 for the sample in the example above.

A systematic sample has the advantage of being quick and easy to use. However care needs to be taken lest there is a cyclic variation in the frame and this is picked up in the sample because it too is organised cyclically.

t

tally, tally chart a method of keeping count using blocks of five.

|	= 1
||	= 2
|||	= 3
||||	= 4
~~||||~~	= 5
~~||||~~ |	= 6
~~||||~~ ~~||||~~ ||	= 12

target population the population from which it is desired to select a sample.

t distribution the distribution t_n defined by $t_n = \dfrac{N(0,1)}{\sqrt{\dfrac{\chi_n^2}{n}}}$.

The *t* distribution underlies the *t* test, one of the most commonly used hypothesis tests, and it is frequently used in calculating confidence intervals. However the form given above is essentially of theoretical interest since it is usual to work with tabulated values.

The distribution of the means of samples of size *n*, drawn from a Normal population with mean μ and standard deviation σ, is itself Normal with mean μ and standard deviation $\dfrac{\sigma}{\sqrt{n}}$, whatever the size of the sample. Often, however, the population standard deviation, σ, is unknown, and can only be estimated from the sample, using

$$s = \sqrt{\frac{1}{n-1}S_{xx}} \text{ where } S_{xx} = \sum_{i=1}^{n}(x_i - \bar{x})^2 .$$

If this estimate, *s*, is used instead of σ, so that the standardised variable is $\dfrac{(\bar{x}-\mu)}{\dfrac{s}{\sqrt{n}}}$ rather than $\dfrac{(\bar{x}-\mu)}{\dfrac{\sigma}{\sqrt{n}}}$, the distribution of the means is t_{n-1} rather than N(0,1).

The t distribution is different for each value of n. As n increases, it approaches the Normal distribution ever more closely, as can be seen in Fig. 120.

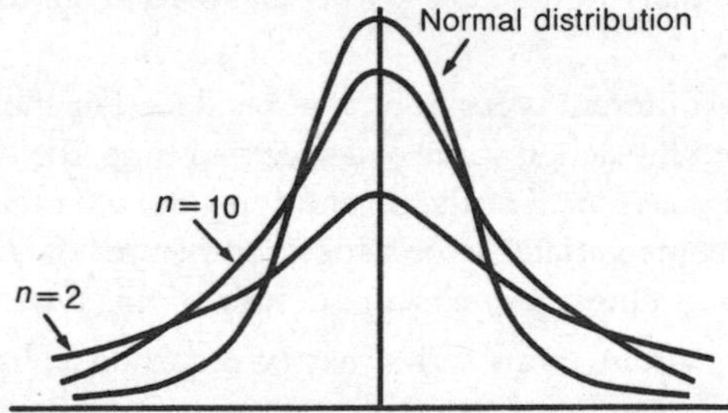

Fig. 120. **t distribution.** Graph showing t distributions ($n = 2$, and $n = 10$) and the Normal distribution.

Tables of the t distribution are usually given only for critical values of t (e.g. 10%, 5%, 2%, 1%, 0.2%, 0.1%). To give full tables for a reasonable choice of the degrees of freedom would be very cumbersome.

test statistic a statistic which is calculated from sample data in order to test a hypothesis about the underlying population or populations. Examples are X^2 for the χ^2 test and t for the t test.

Theil's incomplete method, Theil's complete method. See RESISTANT LINE.

three-quarters high rule an empirical rule for ensuring that a bar chart or vertical line chart gives the right impression. It states that the maximum height should be three-quarters of the horizontal displacement.

In the vertical line chart in Fig. 121, the height, 2.4 cm, is $^3/_4$ of the horizontal distance, 3.2 cm.

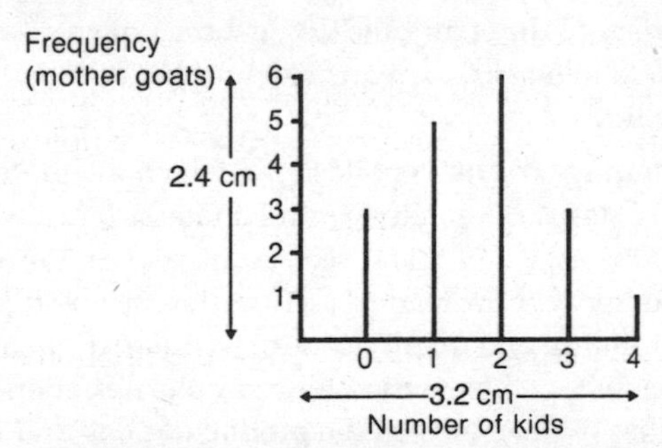

Fig. 121. **Three-quarters high rule.**

time series a set of values of a variable recorded over a period of time. Examples are the oxygen level at a point in the river every hour and the mean daily rainfall at a particular weather station taken over every month.

There are two different types of time series data. For the first type, the data record the actual value at a specified time; the oxygen levels come into this category. For the second type, the data record the aggregate of some variable over a specified period; the mean daily rainfall at the weather station is an example of this type.

The variation within a time series may be regarded as having four components; T, trend; S, seasonal component; C, cyclic component and V, irregular variation.

In an *additive model*, the value of a particular term is given by:

$$X = X_0 + T + S + C + V$$

where X_0 is a base value.

In a *multiplicative model*, it is given by:

$$X = X_0 \times T \times S \times C \times V$$

The effect of seasonal and cyclic components can be eliminated by taking suitable moving averages. For an additive model, the values of the variable are averaged by taking the arithmetic mean; for a multiplicative model, the averaging is done by taking the geometric mean.

A graph showing a time series is sometimes called a *historigram*.

Time series are often used to forecast future values.

See MOVING AVERAGE, RUNNING MEDIANS.

time series analysis statistical analysis of data taken from a single source at a variety of different times. By contrast in cross-section analysis, the data to be analysed are taken from a number of sources, all at the same time.

total quality management a philosophy which integrates the methodology of statistical quality control and assurance with management. One aspect of this is seeing one part of a firm as a supplier to another part. Instead of each section just being concerned about its performance and quality control techniques, an attempt is made to link these together so that, for example, designers are aware of the capabilities of machines on the production line and hence the effect of their tolerance limits.

transect see QUADRAT.

transition matrix a matrix relating the probabilities of a stochastic process being in a particular state to its previous state. It is used to model a stochastic process. A transition matrix is associated with every Markov chain.

There are two common conventions. In one the probability distribution is written as a row matrix, in the other as a column matrix. The example below is presented twice, once with each convention.

Example: At a certain place, if it is fine on one day, the probability of its being fine on the next day is $\frac{2}{3}$; if, however, it is wet, the probability of the next day being fine is $\frac{1}{2}$.

One week, Monday is fine; what is it probability of Wednesday being fine?

Using the convention that the probability distribution is expressed as a row matrix.

The information on probabilities can be set out in table form:

		TOMORROW'S WEATHER	
		Fine	**Wet**
TODAY'S WEATHER	**Fine**	$\frac{2}{3}$	$\frac{1}{3}$
	Wet	$\frac{1}{2}$	$\frac{1}{2}$

This can be written as the transition matrix,

$$\begin{pmatrix} \frac{2}{3} & \frac{1}{3} \\ \frac{1}{2} & \frac{1}{2} \end{pmatrix}$$

The fact that Monday is fine means that Monday's probability distribution is:

$$\underset{\text{Fine}}{(1} \quad \underset{\text{Wet}}{0)}$$

The probability distribution for Tuesday is given by:

Monday's probability distribution	×	Transition matrix	=	Tuesday's probability distribution
$(1 \quad 0)$	×	$\begin{pmatrix} \frac{2}{3} & \frac{1}{3} \\ \frac{1}{2} & \frac{1}{2} \end{pmatrix}$	=	$\underset{\text{Fine}}{(\frac{2}{3}} \quad \underset{\text{Wet}}{\frac{1}{3})}$

Similarly

Tuesday's probability distribution	×	Transition matrix	=	Wednesday's probability distribution
$\begin{pmatrix}\frac{2}{3} & \frac{1}{3}\end{pmatrix}$	×	$\begin{pmatrix}\frac{2}{3} & \frac{1}{3} \\ \frac{1}{2} & \frac{1}{2}\end{pmatrix}$	=	$\begin{pmatrix}\frac{11}{18} & \frac{7}{18}\end{pmatrix}$ Fine Wet

In this convention, the numbers in each row of a transition matrix always add to 1. In this case

$$\frac{2}{3}+\frac{1}{3}=1 \quad \text{and} \quad \frac{1}{2}+\frac{1}{2}=1$$

Using the convention that the probability distribution is expressed as a column matrix.

In this case the table giving information on probabilities is

		TODAY'S WEATHER Fine	Wet
TOMORROW'S WEATHER	**Fine**	$\frac{2}{3}$	$\frac{1}{2}$
	Wet	$\frac{1}{3}$	$\frac{1}{2}$

Thus the transition matrix is

$$\begin{pmatrix}\frac{2}{3} & \frac{1}{2} \\ \frac{1}{3} & \frac{1}{2}\end{pmatrix}$$

and Monday's probability distribution is written as $\begin{pmatrix}1 \\ 0\end{pmatrix}\begin{matrix}\text{Fine} \\ \text{Wet}\end{matrix}$

The probability distribution for Tuesday is given by:

$$\begin{pmatrix}\frac{2}{3} & \frac{1}{2} \\ \frac{1}{3} & \frac{1}{2}\end{pmatrix} \times \begin{pmatrix}1 \\ 0\end{pmatrix} = \begin{pmatrix}\frac{2}{3} \\ \frac{1}{3}\end{pmatrix}\begin{matrix}\text{Fine} \\ \text{Wet}\end{matrix}$$

Similarly the probability distribution for Wednesday is

$$\begin{pmatrix}\frac{2}{3} & \frac{1}{2} \\ \frac{1}{3} & \frac{1}{2}\end{pmatrix} \times \begin{pmatrix}\frac{2}{3} \\ \frac{1}{3}\end{pmatrix} = \begin{pmatrix}\frac{11}{18} \\ \frac{7}{18}\end{pmatrix}\begin{matrix}\text{Fine} \\ \text{Wet}\end{matrix}$$

With this convention it is the numbers in each column of the transition matrix that always sum to 1.

Transition matrices need not be 2×2. If instead of defining the weather on any day to be wet or fine, as in the example, three possibilities, fine, mixed or wet weather had been allowed, then a 3×3 transition matrix would have been needed.

Sometimes a process, represented by a transition matrix **T**, settles down to a steady equilibrium state in which the probability distribution, **p**, is the same from one time interval to the next. In such cases the probability distribution is given by

$$\mathbf{p} = \mathbf{pT} \text{ or } \mathbf{p} = \mathbf{Tp}$$

according to which convention is being used.

The probability distribution, p, n time intervals after an initial probability distribution p_0 is given by

$$\mathbf{p} = \mathbf{p}_0\mathbf{T}^n \text{ or } \mathbf{p} = \mathbf{T}^n\mathbf{p}_0.$$

The subsequent steps are illustrated for the convention where the probability distribution is written as a column matrix. It should, however, be noted that an alternative to this technique is to use a calculator or computer repeatedly.

The matrix $\mathbf{T}^n$ can be simplified by writing T in the form

$$\mathbf{T} = \mathbf{QDQ}^{-1}$$

where **D** is the matrix with the *eigenvalues* of **T** in the leading diagonal and zeros elsewhere, and **Q** is the matrix consisting of the corresponding *eigenvectors*. (See the end of this entry for a note on how to calculate eigenvalues and eigenvectors). Thus, the transition matrix in the example above can be written as:

$$\begin{matrix} \mathbf{T} & = & \mathbf{Q} & \mathbf{D} & \mathbf{Q}^{-1} \end{matrix}$$

$$\begin{pmatrix} \frac{2}{3} & \frac{1}{2} \\ \frac{1}{3} & \frac{1}{2} \end{pmatrix} = \begin{pmatrix} 3 & 1 \\ 2 & -1 \end{pmatrix} \begin{pmatrix} 1 & 0 \\ 0 & \frac{1}{6} \end{pmatrix} \begin{pmatrix} \frac{1}{5} & \frac{1}{5} \\ \frac{2}{5} & -\frac{3}{5} \end{pmatrix}$$

where 1 and $\frac{1}{6}$ are the eigenvalues of T, and $\begin{pmatrix} 3 \\ 2 \end{pmatrix}$ and $\begin{pmatrix} 1 \\ -1 \end{pmatrix}$ the corresponding eigenvectors.

One of the eigenvalues of a transition matrix is always 1.

Writing a transition matrix in this way allows it to be raised to any power, n, using the formula

$$\mathbf{T}^n = \mathbf{QD}^n\mathbf{Q}^{-1}$$

Since the matrix **D** is in diagonal form, it is easily raised to the power n; for example

$$\begin{pmatrix} 1 & 0 \\ 0 & \frac{1}{6} \end{pmatrix}^n = \begin{pmatrix} 1^n & 0 \\ 0 & (\frac{1}{6})^n \end{pmatrix} = \begin{pmatrix} 1 & 0 \\ 0 & (\frac{1}{6})^n \end{pmatrix}$$

Example: if it is fine one Monday, what is the probability distribution for (a) the following Monday, (b) the same day a year later?

(a) The probability is given by:

$$\begin{pmatrix} \frac{2}{3} & \frac{1}{2} \\ \frac{1}{3} & \frac{1}{2} \end{pmatrix}^7 \begin{pmatrix} 1 \\ 0 \end{pmatrix}$$

$$= \begin{pmatrix} 3 & 1 \\ 2 & -1 \end{pmatrix} \begin{pmatrix} 1 & 0 \\ 0 & \frac{1}{6} \end{pmatrix}^7 \begin{pmatrix} \frac{1}{5} & \frac{1}{5} \\ \frac{2}{5} & -\frac{3}{5} \end{pmatrix} \begin{pmatrix} 1 \\ 0 \end{pmatrix}$$

$$= \begin{pmatrix} 3 & 1 \\ 2 & -1 \end{pmatrix} \begin{pmatrix} 1 & 0 \\ 0 & 1/279936 \end{pmatrix} \begin{pmatrix} \frac{1}{5} & \frac{1}{5} \\ \frac{2}{5} & -\frac{3}{5} \end{pmatrix} \begin{pmatrix} 1 \\ 0 \end{pmatrix}$$

$$= \begin{pmatrix} 0.600001 & 0.599998 \\ 0.399999 & 0.400002 \end{pmatrix} \begin{pmatrix} 1 \\ 0 \end{pmatrix}$$

$$= \begin{pmatrix} 0.600001 \\ 0.399999 \end{pmatrix} \begin{matrix} \text{Fine} \\ \text{Wet} \end{matrix}$$

(b) The same day next year, 365 days later, the probability distribution is given by:

$$\begin{pmatrix} \frac{2}{3} & \frac{1}{2} \\ \frac{1}{3} & \frac{1}{2} \end{pmatrix}^{365} \begin{pmatrix} 1 \\ 0 \end{pmatrix}$$

$$\begin{pmatrix} \frac{2}{3} & \frac{1}{2} \\ \frac{1}{3} & \frac{1}{2} \end{pmatrix}^{365} \begin{pmatrix} 1 \\ 0 \end{pmatrix} = \begin{pmatrix} 3 & 1 \\ 2 & -1 \end{pmatrix} \begin{pmatrix} 1 & 0 \\ 0 & \frac{1}{6} \end{pmatrix}^{365} \begin{pmatrix} \frac{1}{5} & \frac{1}{5} \\ \frac{2}{5} & -\frac{3}{5} \end{pmatrix} \begin{pmatrix} 1 \\ 0 \end{pmatrix}$$

$$= \begin{pmatrix} 3 & 1 \\ 2 & -1 \end{pmatrix} \begin{pmatrix} 1 & 0 \\ 0 & 0 \end{pmatrix} \begin{pmatrix} \frac{1}{5} & \frac{1}{5} \\ \frac{2}{5} & -\frac{3}{5} \end{pmatrix} \begin{pmatrix} 1 \\ 0 \end{pmatrix}$$

$$= \begin{pmatrix} \frac{3}{5} \\ \frac{2}{5} \end{pmatrix} \begin{matrix} \text{Fine} \\ \text{Wet} \end{matrix}$$

(This case is, in effect, one where $n = \infty$. The result in such a case gives the unconditional probability distribution.) Thus, in this example, the probability of a day one year hence being fine is $\frac{3}{5}$, wet $\frac{2}{5}$.

Note: the eigenvalues λ_1, and λ_2, of a matrix $\begin{pmatrix} a & c \\ b & d \end{pmatrix}$ (where a, b, c and d are known numbers) is found by solving the equation $(a - \lambda)(d - \lambda) - bc = 0$. The eigenvector corresponding to λ_1 is found by solving $\begin{pmatrix} a & c \\ b & d \end{pmatrix}\begin{pmatrix} x \\ y \end{pmatrix} = \lambda_1 \begin{pmatrix} x \\ y \end{pmatrix}$; similarly for that corresponding to λ_2.

transition probability see STOCHASTIC PROCESS.

tree diagram a diagram in which events or relationships are represented by points joined by lines like the branches of a tree lying on its side. It is laid out so that the rules, 'multiply along the branches' and 'add vertically' apply for probabilities.

Example: For a particular type of chicken, $\frac{3}{5}$ of the eggs contain male chicks and $\frac{2}{5}$ female. The probability of a male hatching and surviving to adulthood is $\frac{3}{4}$, that for a female $\frac{1}{2}$.

A poultry farmer incubates 100 eggs. How many adults can he expect? How many will be female?

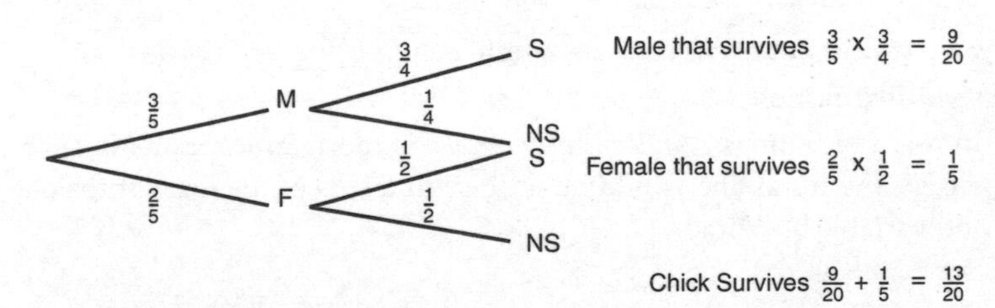

Fig. 122. **Tree diagram.** M male, F female, S survives, NS does not survive.

The probability that an egg ends up as an adult is $\frac{13}{20}$.

The expected number of adults is $\frac{13}{20} \times 100 = 65$.

The probability that an egg ends up as an adult female is $\frac{1}{5}$.

The expected number of females is $\frac{1}{5} \times 100 = 20$.

trend the underlying direction and rate of change in a time series, when allowance has been made for the effects of seasonal components, cyclic components, and irregular variation. The equation of a trend may be called the trend line; it need not be straight although it often is (Fig. 123).

Example: the UK truffle production from 2000 to 2030 is as follows:

Year	2000	2010	2020	2030
Truffle yield (kg)	230	270	310	350

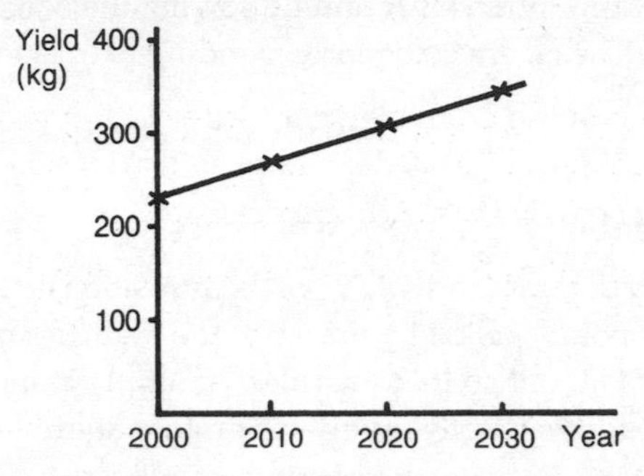

Fig. 123. **Trend.** The straight trend line for production of truffles from 2000 to 2030.

It is easy to see that these figures fit the trend line,

$$\text{Yield} = 230 + 4t$$

where t is the number of years since 2000. In this case, the yield was found by an *additive model*.

One would not expect real data to fit a trend line exactly, as in the example.

In many situations a *multiplicative model* is more appropriate; in such a case, the trend line would be written in terms of the logarithms of the variable and time.

See TIME SERIES.

trial **1.** a general term covering certain types of experiment, for example clinical trials in the pharmaceutical industry.

2. A Bernoulli trial is an experiment with fixed probability p of success, $1 - p$ of failure. In a sequence of Bernoulli trials, the probability of success remains constant from one trial to the next, independent of the outcome of previous trials.

Examples:

Tossing a coin: success is getting a head.

Throwing a die: success is getting a 6.

Selecting a housewife at random: success is if she uses a particular brand of soap.

Selecting a pepper moth: success is if it is the dark variety.

It is essential that success be clearly defined so that a trial is either successful or unsuccessful. Trials are said to be independent if the probability of success in one is not influenced by the outcome of any of the others.

The distribution of the probabilities of different numbers of successes (0 to n) in n Bernoulli trials is the binomial distribution, also occasionally called the Bernoulli distribution.

The geometric distribution gives the probability that the first success will occur at the nth Bernoulli trial, where n takes values from 1 to infinity.

Pascal's distribution, (or the negative binomial distribution) gives the probability that the rth success occurs at the nth Bernoulli trial, where n takes values from r to infinity.

t test or **Student's t test** a test used on the means of small random samples of one of these null hypotheses:

(a) That the sample has been drawn from a population with a given mean.

(b) That two samples have both been drawn from populations with equal means.

For the single sample test, it is assumed that the sample has been drawn from a Normal population. The standard deviation is unknown and so is estimated from the sample data, using

$$s = \sqrt{\frac{1}{n-1} S_{xx}} \text{ where } S_{xx} = \sum_{i=1}^{n} (x_i - \bar{x})^2 .$$

For the two sample test, there is an equivalent assumption that both samples have been drawn from Normal populations. There is also an assumption that the standard deviation (or variance) of the two populations is the same; this is estimated from the sample data.

Example (one sample): the annual rainfall at a weather station has been found over many years to be Normally distributed with a mean value of 120 cm. The rainfall for 10 years, 2006-2015, has been 130,

151, 109, 138, 156, 145, 101, 129, 138, and 103 cm. Is there evidence, at the 5% significance level, of any change in annual rainfall?

H_0: $\mu = 120$ the mean rainfall is unaltered.

H_1: $\mu \neq 120$ the mean rainfall has changed.

5% significance level

2-tail test

Since the sample is small, and the population standard deviation is unknown, a t test must be used.

Sample size, $n = 10$

Sample mean, $\overline{x} = 130$

Estimated population standard deviation, $= 19.67$

The statistic t is given by:

$$t = \frac{\overline{x} - \mu}{\frac{s}{\sqrt{n}}} = \frac{130 - 120}{\frac{19.67}{\sqrt{10}}} = 1.608$$

There are 10-1 = 9 degrees of freedom.

For 9 degrees of freedom (see Table 3) the critical value for t at the 5% level is 2.262.

Since $1.608 < 2.262$, the null hypothesis is accepted at the 5% significance level; the evidence does not support the claim that the mean rainfall has changed.

Example (two samples): a geologist has two samples of ore.

Sample 1:	$n_1 = 7$	$\overline{x}_1 = 81$(g)	$s_1 = 8$
Sample 2:	$n_2 = 10$	$\overline{x}_2 = 72$(g)	$s_2 = 6$

He wishes to test, at the 5% significance level, whether the two samples have been drawn from the same source.

H_0: $\mu_1 = \mu_2$. The two samples are taken from populations with equal means.

H_1: $\mu_1 \neq \mu_2$. They are drawn from populations with different means.

5% significance level

2-tail test

If the null hypothesis is true, the sampling distribution of the difference of the means of the two samples has:

$$Mean = 0 \quad Standard\ deviation = \sqrt{\frac{\sigma^2}{n_1} + \frac{\sigma^2}{n_2}}$$

where n_1 and n_2 are the sizes of the two samples. The usual estimator for σ, the assumed common population standard deviation, from two samples is the pooled estimator:

$$\sqrt{\frac{(n_1 - 1)s_1^2 + (n_2 - 1)s_2^2}{(n_1 + n_2 - 2)}}$$

The degrees of freedom are $v = n_1 + n_2 - 2$. In this example, the estimated population standard deviation is:

$$\sqrt{\frac{(6 \times 8^2 + 9 \times 6^2)}{(7 + 10 - 2)}} = 6.87$$

and so the standard deviation of the distribution is

$$\sqrt{\left(\frac{6.87^2}{7} + \frac{6.87^2}{10}\right)} = 3.39$$

The difference in the two sample means is 81-72 = 9, so

$$t = \frac{9 - 0}{3.39} = 2.66$$

The critical value of t for $v = 15$ (= 7 + 10 – 2) at the 5% significance level is 2.131.

Since 2.66 > 2.131 the null hypothesis is rejected at this level; the geologist would not be justified in thinking the two samples came from the same source.

This test assumes that the variances of the two populations are equal. In this example, that would probably be the case if the two samples of ore were drawn from the same source.

Paired or matched samples are frequently small in size and so the t test is often used on them. See PAIRED SAMPLE TEST, TWO-SAMPLE TEST.

two-sample test a test on some aspect of the difference between two independent populations, given a random sample from each. A common example is to test whether the population means are equal. Other examples are to test whether the population variances are equal, or to test whether a particular proportion is the same in both populations.

The example which follows illustrates one type of two sample test: a Normal test on the means of 2 samples from populations with a common variance.

Example: a study is being carried out into a particular type of mouse. Earlier work suggests that the standard deviation of the weight of adult males is 4.4 g. A sample of 100 mice is collected in England and has mean weight 21.52 g. Another sample of 150 mice is collected in Scotland and their mean weight is 22.60 g. Is there evidence at the 5% significance level to suggest that the mean of the population weights is different in England and Scotland?

H_0: $\mu_1 = \mu_2$, there is no difference in the population means.

H_1: $\mu_1 \neq \mu_2$, there is a difference in the population means.

5% significance level

2-tail test

The test statistic for this type of test is $z = \dfrac{(\bar{x}_1 - \bar{x}_2) - (\mu_1 - \mu_2)}{\sigma\sqrt{\dfrac{1}{n_1} + \dfrac{1}{n_2}}}$.

In this case: $\bar{x}_1 = 21.52$, $n_1 = 100$ (English sample)

$\bar{x}_2 = 22.60$, $n_2 = 150$ (Scottish sample)

$\sigma = 4.4$

$\mu_1 = \mu_2$, according to the null hypothesis

So the test statistic is $z = \dfrac{(21.52 - 22.60) - (0)}{4.4\sqrt{\dfrac{1}{100} + \dfrac{1}{150}}} = -1.90$

The critical values for z are ± 1.96.

Since $-1.90 > -1.96$, the null hypothesis is accepted. The evidence does not support the view that there is a difference in the mean weights. However with a result as close as this it might be prudent to carry out further sampling.

Examples of two-sample tests include the following:

Test	H_0	Test Statistic
Normal test on the means of 2 samples from populations with a common variance.	$\mu_1 = \mu_2$ or $\mu_1 - \mu_2 =$ some fixed value	$\dfrac{(\bar{x}_1 - \bar{x}_2) - (\mu_1 - \mu_2)}{\sigma\sqrt{\dfrac{1}{n_1} + \dfrac{1}{n_2}}}$
Normal test on the means of 2 samples from populations with different variance.	$\mu_1 = \mu_2$ or $\mu_1 - \mu_2 =$ some fixed value	$\dfrac{(\bar{x}_1 - \bar{x}_2) - (\mu_1 - \mu_2)}{\sqrt{\left(\dfrac{\sigma_1^2}{n_1} + \dfrac{\sigma_2^2}{n_2}\right)}}$
t test for the difference in the means of 2 samples from populations with a common variance. (The populations must be Normally distributed.)	$\mu_1 = \mu_2$ or $\mu_1 - \mu_2 =$ some fixed value	$\dfrac{(\bar{x}_1 - \bar{x}_2) - (\mu_1 - \mu_2)}{s\sqrt{\dfrac{1}{n_1} + \dfrac{1}{n_2}}}$ where $s^2 = \dfrac{(n_1 - 1)s_1^2 + (n_2 - 1)s_2^2}{(n_1 + n_2 - 2)}$
Wilcoxon 2-sample Rank Sum Test or Mann-Whitney U Test.	The two samples are drawn from a common distribution.	W = Sum of ranks of smaller samples or U is the smaller of $= n_1 n_2 + \dfrac{n_1(n_1+1)}{2} - R_1$ and $= n_1 n_2 + \dfrac{n_2(n_2+1)}{2} - R_2$
F test on ratio of two variances	$\sigma_1^2 = \sigma_2^2$	$\dfrac{s_1^2}{s_2^2}$ (where $s_1^2 > s_2^2$)

two-tail test see ONE- AND TWO-TAIL TESTS.

two-way analysis of variance see ANALYSIS OF VARIANCE, FREIDMAN'S TWO-WAY ANALYSIS OF VARIANCE BY RANK.

type 1 and type 2 errors those errors which occur when, respectively, a true null hypothesis is rejected and a false null hypothesis is accepted.

Example: It is proposed that coins to be used in a casino are tested for bias by being tossed six times. If a coin comes either heads or tails all six times it is rejected.

This is, in effect, a test of

H_0: The coin is not biased: $p = \frac{1}{2}$.

H_1: The coin is biased: $p \neq \frac{1}{2}$.

(a) What is the probability that a good coin is rejected (Type 1 error)?

The probability of a head = probability of a tail = $\frac{1}{2}$. Thus the probability of 6 heads is $(\frac{1}{2})^6$ and similarly the probability of 6 tails is $(\frac{1}{2})^6$.

So the probability of a good coin being rejected, a Type 1 error, is:

$$\left(\tfrac{1}{2}\right)^6 + \left(\tfrac{1}{2}\right)^6 = \tfrac{1}{32}$$

This is the significance level of the test.

(b) A coin is in fact biased with probability $^1/_4$ of coming tails. What is the probability that it is accepted (Type 2 error)?

Probability of 6 heads = $(^3/_4)^6$

Probability of 6 tails = $(^1/_4)^6$

Sp the probability of the coin being accepted (a Type 2 error) is:

$$1 - \left(\tfrac{3}{4}\right)^6 - \left(\tfrac{1}{4}\right)^6 = 0.822$$

In this example, the probability of a Type 1 error is quite small, of a Type 2 error much larger. The probability of a Type 2 error could be reduced by making the test less severe (i.e., fewer than 6 losses) but this would increase the probability of a Type 1 error. In this particular case, a Type 1 error (rejecting a good coin) would not matter much, but a Type 2 error (accepting a biased coin) could be serious. Consequently, the proposed test is quite unsuitable.

A better test would be to toss the coin 6 times, and accept it only if the results were 3 heads and 3 tails. In that case, the probability of a Type 1 error, rejecting the unbiased coin, is 0.6875. The probability of a Type 2 error, accepted a biased coin, is 0.132 in the case when the probability of it coming tails is $^1/_4$. A more severe test (e.g. tossing coins 20 times and accepting only those that come 10 heads and 10 tails) would result in a higher number of unbiased coins being rejected, but fewer biased ones being accepted. See SIGNIFICANCE LEVEL.

Typical Year Index see INDEX NUMBER.

U

unbiased estimator see BIASED ESTIMATOR.

uniform distribution the discrete distribution in which the probability of each possible value of the variable is the same.

Example: the distribution of the score from throwing a fair single die is uniform, with each of thc values 1, 2, 3, 4, 5 and 6 having the same probability of $^1/_6$. See Fig. 124.

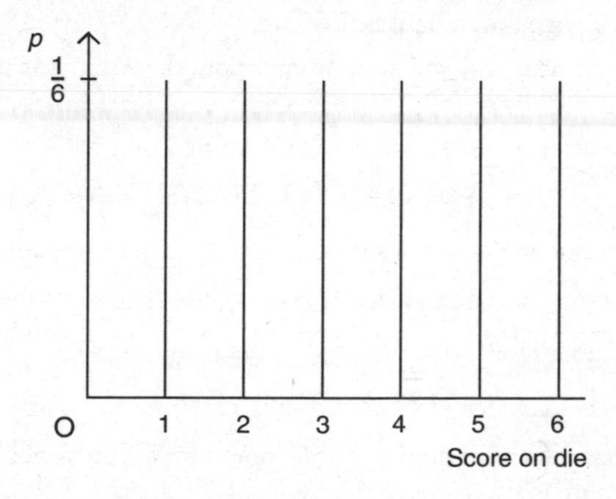

Fig. 124. **Uniform distribution.**

The term uniform distribution is also sometimes applied to a continuous distribution whose probability distribution is constant between certain limits and zero elsewhere. Another name for this distribution is the *rectangular distribution*.

The terms discrete uniform distribution and continuous uniform distribution are sometimes used to distinguish between the two types of uniform distribution.

union see SET.

universal set see SET.

V

variable error error occurring in certain types of experiments on living subjects due to factors which vary during the course of an experiment, such as tiredness, or experience.

variable testing testing a product for quality control when the result is a quantitative measurement like the breaking strain of a fishing line. Variable testing may be contrasted with attribute testing where the result is a verdict of either 'good' or 'defective' as, for example, in a check on a light bulb.

variance a measure of dispersion or variation, the square of standard deviation.

Population variance, notation Var(X), is defined by

$$\mathrm{Var}(X) = \mathrm{E}\left[\left(X - \mathrm{E}(X)\right)^2\right] \quad \text{(See EXPECTATION).}$$

For a continuous variable with probability density function f(x)

$$\mathrm{Var}(X) = \int x^2 \mathrm{f}(x)\,\mathrm{d}x - [\mathrm{E}(X)]^2$$

For a discrete variable with probability distribution p(x_i)

$$\mathrm{Var}(X) = \sum_i x_i^2 \mathrm{p}(x_i) - [\mathrm{E}(X)]^2$$

Sample variance is a random variable, notation S^2; the notation for a particular value is s^2.

The following formulae are commonly used for calculating sample variance. (See SUM OF SQUARES).

Unsorted data with n values: $x_1, x_2, \ldots x_n$.

$$s^2 = \frac{S_{xx}}{n-1} \text{ where } S_{xx} = \sum_{i=1}^{n}(x_i - \bar{x})^2 = \sum_{i=1}^{n} x_i^2 - \frac{1}{n}\left(\sum_{i=1}^{n} x_i\right)^2$$

Sorted data with n values in k groups:

Value	x_1	x_2	...	x_k
Frequency	f_1	f_2	...	f_k

$$f_1 + f_2 + \ldots f_k = n$$

$$s^2 = \frac{S_{xx}}{n-1} \text{ where}$$

$$S_{xx} = \sum_{i=1}^{k}([x_i - \bar{x}]^2 f_i) = \sum_{i=1}^{k}(x_i^2 f_i) - \frac{1}{n}\left(\sum_{i=1}^{k} x_i f_i\right)^2$$

However a divisor n is also sometimes used in these formulae, rather than $n - 1$.

For variables X and Y, and constants a, b and c.

$$\text{Var}(aX) = a^2\text{Var}(X)$$

$$\text{Var}(X + c) = \text{Var}(X)$$

$$\text{Var}(X \pm Y) = \text{Var}(X) + \text{Var}(Y) \pm 2\text{cov}(X, Y)$$

$$\text{Var}(aX \pm bY) = a^2\text{Var}(X) + b^2\text{Var}(Y) \pm 2ab \text{ cov}(X, Y)$$

Variance is used extensively in statistics. Although it is not easy to give a conceptual meaning to variance, it is important because it occurs frequently in theoretical work.

variate **1.** a random variable.

2. a numerical value taken by a random variable.

variation dispersion or spread, usually about some value. For example, variation about the mean is called deviation.

Venn diagram or **Euler diagram** see SET.

vertical line chart a method of displaying information similar to a bar chart but using lines instead of bars (see Fig. 125 overleaf). A vertical line chart is usually used for displaying numerical data. Compare BAR CHART.

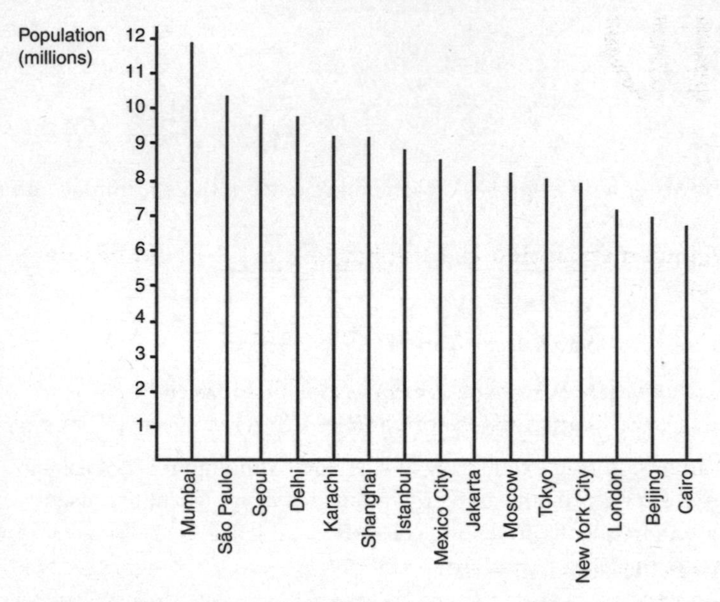

Fig. 125. **Vertical line chart.** The world's largest cities. (Note; figures for city sizes vary considerably with the source; this is due in part to different definitions of city boundaries.

Weibull distribution the distribution with probability density function:

$$f(x) = \alpha\beta x^{\beta-1}e^{-\alpha x^{\beta}}$$

$(x > 0)$ where α, β are parameters of the distribution. If $\beta = 1$, this is the exponential distribution.

This distribution is often used in reliability theory (among other applications).

weight one of a set of numbers used to multiply the values of particular data items when combining them into an aggregate value. Items of greater relative importance may be given greater weight. See WEIGHTED MEAN.

weighted mean an average value that takes account of the relative importance of the constituent elements. The weighted mean is given by:

$$\frac{w_1x_1 + w_2x_2 + \ldots + w_nx_n}{w_1 + w_2 + \ldots + w_n}$$

where w_1, w_2, ..., w_n are the weights given to the values x_1, x_2, ... x_n.

Example: a railway company conducts regular surveys into passenger satisfaction. The passengers rate their experience on a scale of 0 (bad) to 10 (good). The company works out an average value but weights the responses according to whether passengers are season ticket holders, regular users or occasional users, using the ratio 6: 3:1. (In this case the numbers 6, 3 and 1 are the *weights*.)

Thus their weighted mean is

$$\frac{6x_1 + 3x_2 + x_3}{10}$$

where x_1, x_2, and x_3 are the figures for the three groups.

One month the survey gives

$x_1 = 8$ (season ticket holders)

$x_2 = 6$ (regular users)

$x_3 = 4$ (occasional users)

So the weighted mean is

$$\frac{6\times8+3\times6+1\times4}{10} = 7.$$

In this example the views of season ticket holders are particularly heavily weighted.

Wilcoxon tests a set of non-parametric tests for the population median, or for comparing medians.

1. Wilcoxon single sample test

This test is illustrated in the following example.

Example: an opinion pollster asks 12 people their views on a proposed bypass. They choose one out of 7 possible responses which are recorded on a 1 to 7 scale.

Response	Score
Very strongly opposed	1
Quite strongly opposed	2
Mildly opposed	3
Don't mind	4
Mildly support	5
Quite strongly support	6
Very strongly support	7

Their responses are as follows: 1, 6, 5, 4, 3, 4, 2, 7, 4, 7, 5, 4.

Can these responses be taken as evidence, at the 5% significance level, that most people support the bypass?

H_0: The median response is 4.

H_1: The median response is greater than 4 (so most people support the bypass).

5% significance level

1-tail test

The procedure is shown in the table below.

Score, x	$x-4$	$\|x-4\|$	Rank	Signed rank	Negative Ranks
1	–3	3	7	–7	(–)7
6	2	2	$4\frac{1}{2}$	$4\frac{1}{2}$	
5	1	1	2	2	
4	0	0	–	–	
3	–1	1	2	–2	(–)2
4	0	0	–	–	
2	–2	2	$4\frac{1}{2}$	$-4\frac{1}{2}$	$(-)4\frac{1}{2}$
7	3	3	7	7	
4	0	0	–	–	
7	3	3	7	7	
5	1	1	2	2	
4	0	0	–	–	
			$m = 8$		$W = (-)13\frac{1}{2}$

The 12 values of $|x-4|$, placed in order are:

$$0 \quad 0 \quad 0 \quad 0 \quad 1 \quad 1 \quad 1 \quad 2 \quad 2 \quad 3 \quad 3 \quad 3$$

All cases where the values are 0 are ignored; in this example there are four of them. This reduces the number of values under consideration, denoted by m, from 12 to 8; $m = 8$. They are

$$1 \quad 1 \quad 1 \quad 2 \quad 2 \quad 3 \quad 3 \quad 3.$$

These values are then ranked, starting at the smallest, 1. The three values 1 occupy places 1, 2 and 3 and so they are given the average rank, $\frac{1+2+3}{3} = 2$. Similarly for the remaining values. These ranks are given in the fourth column. Notice that it is the absolute values that have been ranked, so that values of +1 and –1 are ranked equal.

In the fifth column (one from the right), the ranks are given signs. A + sign indicates a positive response (i.e. supporting the bypass), a – sign a negative reponse (opposing the bypass).

The right-hand column contains only the ranks corresponding to negative differences; these are summed to give the total W, in this case $13\frac{1}{2}$.

The critical value of W for $m = 8$ at the 5% significance level for a 1-tail test is 5. (See Table 13).

Since $13\frac{1}{2} \geq 5$, the null hypothesis is accepted.

Notice that in this test the test statistic W must be less than or equal to the critical value for the null hypothesis to be rejected. The reason for this is that the total sum of the ranks is $1+2+ \ldots m = \frac{1}{2}m(m+1)$. In this case the total is $\frac{1}{2}\times 8\times 9 = 36$. The sum of the negative ranks, W, is $13\frac{1}{2}$ and the sum of the positive ranks is $36 - 13\frac{1}{2} = 22\frac{1}{2}$. If the differences in the ranks were divided equally between positive and negative, they would each total $\frac{1}{2}$ of $36 = 18$. The further the rank sums are away from 18, the more likely it is that the median has changed. Taking the negative differences involves looking at the value below the middle and so the smaller the value of W, the more extreme it is.

The procedure can also be used with the positive ranks. In this example the critical value would then be $36–5 = 31$, and so the critical region would be $W \geq 31$.)

If m is not small, W is approximately Normally distributed when the null hypothesis is true, with:

$$Mean = \frac{m(m+1)}{4} \quad \text{and} \quad Variance = \frac{m(m+1)(2m+1)}{24}$$

This can be used to calculate the significance level of the data to a high degree of accuracy; a continuity correction should be used.

2. *Wilcoxon matched-pairs signed-rank test*

The same basic method is used in the Wilcoxon matched-pairs test but in this case it is applied to the differences in ranks of the pairs in a matched sample. The null hypothesis is that the median of the differences is zero. The method is illustrated in the next example.

Example: ten army recruits were given a shooting test. They were then given a lecture on good shooting technique, at the end of which they repeated the test. The results of the two tests, with part of the calculation, are in the table below. Can it be claimed that their performance was better, at the 5% significance level, after the lecture?

H_0: The median of the differences is zero. The lecture had no effect and the underlying population median was the same before and after the lecture.

H_1: The median of the differences is greater than zero. Performance improved after the lecture.

5% significance level

1-tail test

Recruit	*First shoot*	*Second shoot*	*Difference*	*Absolute difference*	*Rank of absolute difference*	*Signed rank of difference*	*Negative ranks*
A	45	46	+1	+1	1½	+1½	
B	32	30	-2	+2	3	-3	(–)3
C	24	30	+6	+6	6	+6	
D	12	12	0	0	-	-	
E	15	19	+4	+4	5	+5	
F	41	40	-1	+1	1½	-1½	(–)1½
G	28	35	+7	+7	7	+7	
H	29	40	+11	+11	9	+9	
I	31	39	+8	+8	8	+8	
J	25	28	+3	+3	4	+4	
					$m = 9$	*Total W*	(–)4½

The *difference* column is the improvement between the first shoot and the second. Having established this, the absolute values of the numbers in the difference column are then ranked, the largest being given the highest rank.

All differences of 0 are ignored; the *D* pair has difference 0. This reduces the number of pairs under consideration, denoted by *m*, from 10 to 9; $m = 9$.

A + sign indicates the difference was positive (i.e., an improvement in shooting), a – sign that it was negative (a worse performance). Notice that it is the absolute values that are ranked, so that values of +1 and –1 tie for the lowest place.

The right-hand column contains only the ranks corresponding to negative differences; these are summed to give the total *W*, in this case (–)$4\frac{1}{2}$.

The critical value (see Table 13) for $m = 9$, for a one-tail test at the 5% significance level is 8.

Since 4.5≤8 the null hypothesis is rejected. The evidence support the alternative hypothesis that the median has increased, and so that the lecture has done some good.

3. *Wilcoxon rank sum test*

This is a two-sample test. The null hypothesis is that the samples are drawn from populations with the same median. The method is illustrated in the following example.

Example: two different research groups are trying different methods for long range weather forecasting. Their success over a given time period is measured on a scale of 0 (no success) to 20 (fully successful). The available data cover 6 periods for group A and 8 for group B.

A: 16, 12, 14, 18, 17, 13

B: 14, 3, 3, 13, 10, 14, 8, 5

Do the data provide evidence, at the 5% significance level, that one group is more successful than the other?

H_0: The two sets of results are drawn from populations with the same median.

H_1: The medians of the underlying populations are different.

5% significance level

2-tail test

The data are ranked as follows

Group A	Rank	Group B
	$1\frac{1}{2}$	3, 3
	3	5
	4	8
	5	10
12	6	
13	$7\frac{1}{2}$	13
14	10	14, 14
16	12	
17	13	
18	14	

The rank sum of group A is $6+7\frac{1}{2}+10+12+13+14 = 62\frac{1}{2}$
The rank sum of group B is $1\frac{1}{2}+1\frac{1}{2}+3+4+5+7\frac{1}{2}+10+10 = 42\frac{1}{2}$

A useful check at this point is provided by the fact that the total rank sum should be $\frac{1}{2}(m+n)(m+n+1)$ where m is the size of the smaller sample (in this case 6), n is the size of the larger sample (in this case 8) and $m+n$ is the total sample size (in this case 6+8 = 14).
So $\frac{1}{2}(m+n)(m+n+1) = \frac{1}{2}\times 14\times 15 = 105$ and $62\frac{1}{2} + 42\frac{1}{2} = 105$.

If the null hypothesis is true, the expected rank sum of the smaller sample is given by:

$$\frac{m}{m+n}\times\frac{1}{2}(m+n)(m+n+1) = \frac{1}{2}m(m+n+1).$$

In this case, this is 45 for group A.

Similarly the expected rank sum of group B is $\frac{1}{2}n(m+n+1) = 60$.

The test statistic, W, is the rank sum of the smaller sample (i.e. group A) and so in this case $W = 62\frac{1}{2}$. (The only reason for using the smaller sample is that the tables are given for $m \le n$.)

The critical value is found in Table 12. In this case $m = 6$ and $n = 8$ and the critical value is 29. However, this is the critical value for the left hand tail in a 2-tail test and in this example it is the right hand tail which is relevant. 29 is 16 less than the expected rank sum of 45 for group A. So the right hand critical value is given by 45+16 = 61. It is this value that is compared with the test statistic $W = 62\frac{1}{2}$.

Since $62^1/_2 \ge 61$, the null hypothesis is rejected. The data provide evidence that there is a difference in the success of the two groups.

This test is very similar to the Mann-Whitney test. The only difference is that the Mann-Whitney test statistic is $\frac{1}{2}m(m+1)$ less than the Wilcoxon (where m is the size of the sample from which the rank sum has been obtained). The Mann-Whitney critical values are correspondingly less.

For larger values of m and n a Normal approximation is often used, with mean and variance as given below.

Wilcoxon form	*Mean* $= \frac{1}{2}mn + \frac{1}{2}m(m+1)$	*Variance* $= \frac{1}{12}mn(m+n+1)$
Mann-Whitney form	*Mean* $= \frac{1}{2}mn$	*Variance* $= \frac{1}{12}mn(m+n+1)$

y

Yates' correction a continuity correction which may be applied when using the chi-squared test on data from a discrete distribution. It is usually used only in tests on 2 × 2 contingency tables with small frequencies. It is applied by reducing the absolute value of (f_o-f_e) by 0.5 in each case. (f_o and f_e are the observed and expected frequency respectively in the four categories.)

Example: a small colony of a previously unknown species of rat is discovered on a remote island. Some of them have spotted fur on their stomachs, others do not; the numbers are shown in the contingency table.

f_o	*Male*	*Female*	*Total*
Spots	6	14	20
No spots	11	9	20
Total	17	23	40

Do the figures give reason, at the 5% SIGNIFICANCE LEVEL, to think that one sex is more likely to have spots than the other?

H_0: The proportions of males and females having spots are the same.

H_1: The proportions are different.

5% significance level

1-tail test

On this hypothesis the expected values are given in the following table:

f_e	*Male*	*Female*	*Total*
Spots	8.5	11.5	20
No spots	8.5	11.5	20
Total	17	23	40

8.5, 11.5 are in the ratio 17:23, and sum to 20.

Using Yates' correction,

$$X^2 = \frac{(6 - 8.5 + .5)^2}{8.5} + \frac{(14 - 11.5 - .5)^2}{11.5} + \frac{(11 - 8.5 - .5)^2}{8.5} + \frac{(9 - 11.5 + .5)^2}{11.5} = 1.64$$

The degrees of freedom, $v = (2 - 1) \times (2 - 1) = 1$.

For $v = 1$, at the 5% significance level, the critical value for $\chi^2 = 3.84$ (see Table 4).

Since $1.64 < 3.84$ the null hypothesis is accepted. The evidence does not support the conclusion that one sex is more likely to have spots than the other, at the 5% level.

Had Yates' correction not been applied, the value of X^2 would have been 2.56 rather than 1.64, but that would not, in this case, have altered the conclusion reached. However, Yates' correction often can make a difference, in which case the results have to be interpreted very carefully.

Yates's correction is conservative. It is always the case that the value of X^2 is smaller with the correction than without it, and so less likely to be significant. While this is comforting in many applications, with a smaller risk of a Type 1 error, there is some concern that the correction, which is only approximate, might be over conservative. See CHI SQUARED TEST.

Yule's coefficient of association see ASSOCIATION.

Z

z transformation see FISHER'S z TRANSFORMATION.

z value the standardised value of a random variable. Standardisation is carried out by subtracting the population mean, and dividing by the standard deviation. Thus the z value is given by:

$$z = \frac{(x - \mu)}{\sigma}$$

The z value is needed when using standardised tables such as those for the Normal distribution. See NORMAL DISTRIBUTION.

Appendixes

Notation

A_1, A_2, A_3	Acceptance numbers
$B(a,b)$	Beta function
$B(n,p)$	Binomial distribution with n trials each with probability p of success
C	Cyclic variation (time series)
c	Class interval
c	Intercept of the straight line $y = mx+c$ with the y-axis
Cov, cov	Covariance
${}^nC_r, \binom{n}{r}$	Binomial coefficient
	The number of ways of choosing r objects from n
D	Test statistic in Kolmorogov-Smirnov tests
E( )	Expectation
e	Base of natural logarithms, 2.71828 (to 5 decimal places)
F	Distribution, test, test statistic
f, f_i	Frequency
f	Sampling fraction
f_o, f_e	Observed and expected frequencies in χ^2 test
f( )	Function
F(x)	Distribution function
f(x)	Probability density function
G(t)	Probability generating function
H	Test statistic for Kruskal-Wallis one-way analysis of variance

H_0	Null hypothesis
H_1	Alternative hypothesis
H_1, H_2	Lower and upper hinges
k	Number of groups
In, $\log_e$	Natural logarithm, logarithm to base e
$\log_{10}$	Logarithm to the base 10
M	Test statistic in Friedman's two-way analysis of variance
m	Sample mean
m	Gradient of the straight line $y = mx + c$
$M_x(t)$	Moment generating function
MS_B, MS_R	Between groups and residual mean square in analysis of variance
N, n	Sample size
N	Population size
N	Number of groups
$N(\mu,\sigma^2)$	Normal distribution with mean μ and standard deviation σ
N(0,1)	Standardised Normal distribution with mean 0 and standard deviation 1
$n(A)$	The number of elements in the set A
p	Probability
	Proportion
p_i	Probability of x_i
$p(x)$	Probability distribution
P( )	Probability
p_0, p_n	Base and current year values
Poisson (λ)	Poisson distribution with parameter λ

q	Probability of non-success in a Bernoulli trial (=1 – p)
q_0, q_n, q_t	Base current and typical year weightings
Q_1, Q_2, Q_3	Lower quartile, median, upper quartile
r	Sample product moment correlation coefficient
r	Integer value of a discrete random variable
r_s	Spearman's rank correlation coefficient
S	Seasonal component (time series)
s	Sample standard deviation
s^2	Sample variance, estimated population variance
S_{xx}, S_{yy}	Sums of squares
S_{xy}	Sum of products
SS_B, SS_R	Between groups and residual sums of squares in analysis of variance
T	Trend (time series)
t	Distribution, test, test statistic
U	Test statistic in Mann-Whitney U test
V	Irregular variation (time series)
Var	Variance
W	Kendall's coefficient of concordance
W	Test statistic in Wilcoxon tests.
$w_1, w_2, \ldots$	Weightings
X	Random variable
X^2	Test statistic for χ^2 and other tests
x	Horizontal axis of cartesian graph
x, x_i	Values of variable

$\overline{x}$	Mean value of x
$\hat{x}$	Estimated value of x
Y	Random variable
y	Vertical axis of cartesian graph
$\overline{y}$	Mean value of y
$\hat{y}$	Estimated value of y
z	Standardised value
	Test statistic for tests using the Normal distribution
	Referring to Fishers z transformation
α	Weighting used in exponential smoothing
α,β	Alternatives for 1 and 2 in Type 1 and Type 2 errors
$\Gamma(\ )$	Gamma function
δx	A small amount of x
ε	Residual
λ	Poisson distribution's parameter
μ	Population mean
$\mu_0, \mu_1, \mu_2, \ldots$	0th, 1st, 2nd, … moments about the mean
$\mu_0', \mu_1', \mu_2', \ldots$	0th, 1st, 2nd, … moments about the origin
v	Degrees of freedom
π	The ratio of the circumference of a circle to its diameter, 3.14159 (to 5 decimal places)
ρ	Population product moment correlation coefficient
Σ	Summation
σ	Population standard deviation
σ^2	Population variance
τ	Kendall's rank correlation coefficient
$\Phi(z)$	Distribution function of standardised Normal distribution
$\varphi(z)$	Probability density function of standardised Normal distribution

χ^2	Distribution, test
$\| \|$	Absolute value, modulus
$\|$	Given (in conditional probability)
!	Factorial
ϵ	Is an element of (in sets)
ε	Universal set
$\varnothing$	Empty set
{ }	Set
$\cap$	Intersection of sets
$\cup$	Union of sets
$\subset$	Is a subset of
$\{\ \}'$	Complement of a set
$\int$	Integral
$\frac{d(\ )}{dx}$	Differential with respect to x
$\frac{\partial(\ )}{\partial x}$	Partial differential with respect to x
$\binom{n}{r}, {}^nC_r$	Binomial coefficient
	The number of ways of choosing r objects from n

Formulae

Probability

Addition law: $P(A \cup B) = P(A) + P(B) - P(A \cap B)$

Multiplication law: $P(A \cap B) = P(A|B).P(B)$

Conditional probability: $P(A|B) = \dfrac{P(A \cap B)}{P(B)}$

Permutations and combinations

Permutations $^{n}P_{r} = \dfrac{n!}{(n-r)!}$

Combinations $^{n}C_{r} = \dbinom{n}{r} = \dfrac{n!}{r!(n-r)!}$

Mean

$$\textit{Arithmetic mean} = \frac{x_1 + x_2 + \ldots x_n}{n}$$

$$\textit{Geometric mean} = \sqrt[n]{x_1 x_2 \ldots x_n}$$

$$\textit{Harmonic mean} = \frac{1}{\frac{1}{n}\left(\frac{1}{x_1} + \frac{1}{x_2} + \ldots + \frac{1}{x_n}\right)}$$

$$\textit{Weighted mean} = \frac{w_1 x_1 + w_2 x_2 + \ldots w_n x_n}{w_1 + w_2 + \ldots + w_n}$$

Sum of squares and products

Squares $S_{xx} = \sum_{i=1}^{n}(x_i - \overline{x})^2 = \sum_{i=1}^{n} x_i^2 - \frac{1}{n}\left(\sum_{i=1}^{n} x_i\right)^2 = \sum_{i=1}^{n} x_i^2 - n\overline{x}^2$

and for grouped data, (where $n = \sum_i f_i$):

$$S_{xx} = \sum_{i=1}^{k}(x_i - \overline{x})^2 f_i = \sum_{i=1}^{k} x_i^2 f_i - \frac{1}{n}\left(\sum_{i=1}^{k} x_i f_i\right)^2 = \sum_{i=1}^{k} x_i^2 f_i - n\overline{x}^2$$

Products $S_{xy} = \sum_{i=1}^{n}(x_i - \overline{x})(y_i - \overline{y}) = \sum_{i=1}^{n} x_i y_i - \frac{1}{n}\sum_{i=1}^{n} x_i y_i = \sum_{i=1}^{n} x_i y_i - n\overline{xy}$

Variance and standard deviation

Standard deviation is the square root of variance

Variance $s^2 = \frac{S_{xx}}{n-1}$ Standard deviation, s = $\sqrt{\frac{S_{xx}}{n-1}}$

A divisor n is sometimes used but in advanced statistics the (n-1) form is almost universal.

Correlation and regression

Population correlation $\rho = \frac{\text{Cov}(X,Y)}{\sigma_X \sigma_Y}$

Sample correlation $r = \frac{S_{xy}}{\sqrt{S_{xx}S_{yy}}}$

Spearman's coefficient of rank correlation $r_s = 1 - \frac{6\sum d^2}{n(n^2 - 1)}$

Kendall's coefficient of rank correlation $\tau = \frac{S}{\frac{1}{2}n(n-1)}$

y on x regression line $y - \overline{y} = \frac{S_{xy}}{S_{xx}}(x - \overline{x})$

Probability distributions

$\sum_{i=1}^{n} \text{p}(x_i) = 1$ $\text{p}(x_i) \geq 0$ all i

Expectation $\text{E}(X) = \sum_{i=1}^{n} x_i \text{p}(x_i)$

Variance $\text{Var}(X) = \sum_{i=1}^{n} (x_i - \mu)^2 \text{p}(x_i)$

Probability density functions

$$\int_{\text{all } X} \mathrm{f}(x)\,\mathrm{d}x = 1 \qquad \mathrm{f}(x) \geq 0 \text{ for all } x.$$

Expectation $\mathrm{E}(X) = \int_{\text{all } X} x\mathrm{f}(x)\,\mathrm{d}x$

Variance $\mathrm{Var}(X) = \int_{\text{all } x} (x-\mu)^2\mathrm{f}(x)\,\mathrm{d}x$

Sampling

From a single sample $(x_1, x_2, \ldots x_n)$

Estimated population mean $= \dfrac{\sum x_i}{n} = \bar{x}$

Estimated population variance $= \dfrac{\sum (x_i - \bar{x})^2}{n-1} = s^2$

From two samples, sizes n_1 and n_2, from the same population

Estimated population mean $= \dfrac{n_1\bar{x}_1 + n_2\bar{x}_2}{n_1 + n_2}$

Estimated population variance $= \dfrac{(n_1 - 1)s_1^2 + (n_2 - 1)s_2^2}{(n_1 + n_2 - 2)}$

For k samples from the same population

Estimated population mean $= \dfrac{n_1\bar{x}_1 + n_2\bar{x}_2 + \ldots + n_k\bar{x}_k}{n_1 + n_2 + \ldots n_k}$

Estimated population variance $= \dfrac{(n_1 - 1)s_1^2 + (n_2 - 1)s_2^2 + \ldots + (n_k - 1)s_k^2}{(n_1 + n_2 + \ldots + n_k - k)}$

Distributions

Discrete distributions

Distribution	Function, $P(X = r)$
Binomial, $B(n,p)$	$\binom{n}{r} q^{n-r} p^r$ where $q = 1 - p$
Poisson (λ)	$e^{-\lambda} \frac{\lambda^r}{r!}$
Discrete uniform	$\frac{1}{n-m+1}$ for $r = m, m+1, \dots n$
Geometric	$q^{r-1} p$ where $q = 1 - p$
Negative binomial	$\binom{r-1}{n-1} q^{r-n} p^n$ where $q = 1 - p$ for $r = n, n+1, \dots$

Continuous distributions

Distribution	p.d.f., $f(x)$
Normal, $N(\mu,\sigma^2)$	$\frac{1}{\sigma\sqrt{2\pi}} e^{-\frac{1}{2}\left(\frac{x-\mu}{\sigma}\right)^2}$
Standardised Normal, $N(0,1)$	$\frac{1}{\sqrt{2\pi}} e^{-\frac{1}{2}z^2}$ where $z = \frac{x-\mu}{\sigma}$
Continuous uniform, or rectangular, on $[a,b]$	$\frac{1}{b-a}$
Exponential	$\lambda e^{-\lambda x}$

Hypothesis tests

Test	Test Statistic
Normal test on a sample mean	$z = \dfrac{(\bar{x} - \mu)}{\dfrac{\sigma}{\sqrt{n}}}$
t test on a sample mean	$z = \dfrac{(\bar{x} - \mu)}{\dfrac{s}{\sqrt{n}}}$
Normal test on the mean of a paired sample	$\dfrac{\bar{d} - \mu_d}{\dfrac{\sigma}{\sqrt{n}}}$
t test on the mean of a paired sample	$\dfrac{\bar{d} - \mu_d}{\dfrac{s_d}{\sqrt{n}}}$
Normal test on the means of 2 samples from populations with a common variance.	$\dfrac{(\bar{x}_1 - \bar{x}_2) - (\mu_1 - \mu_2)}{\sigma\sqrt{\dfrac{1}{n_1} + \dfrac{1}{n_2}}}$
Normal test on the means of 2 samples from populations with different variance.	$\dfrac{(\bar{x}_1 - \bar{x}_2) - (\mu_1 - \mu_2)}{\sqrt{\left(\dfrac{\sigma_1^2}{n_1} + \dfrac{\sigma_2^2}{n_2}\right)}}$
t test for the difference in the means of 2 samples from populations with a common variance. (The populations must be Normally distributed.)	$\dfrac{(\bar{x}_1 - \bar{x}_2) - (\mu_1 - \mu_2)}{s\sqrt{\dfrac{1}{n_1} + \dfrac{1}{n_2}}}$ where $s^2 = \dfrac{(n_1 - 1)s_1^2 + (n_2 - 1)s_2^2}{(n_1 + n_2 - 2)}$
χ^2 test for variance	$X^2 = \dfrac{(n-1)s^2}{\sigma^2}$

F test on ratio of two variances	$F = \frac{s_1^2}{s_2^2}$ (where $s_1^2 > s_2^2$)
Test for product moment correlation	$r = \frac{S_{xy}}{\sqrt{S_{xx}S_{yy}}}$
Spearman's test for rank correlation	$r_s = 1 - \frac{6\sum d^2}{n(n^2-1)}$
Kendall's test for rank correlation	$\tau = \frac{S}{\frac{1}{2}n(n-1)}$
χ^2 test for goodness of fit	$X^2 = \sum_{\text{all groups}} \frac{(f_o - f_e)^2}{f_e}$
Kolmogorov-Smirnov test for goodness of fit	*D* = Largest difference between cumulative frequency distributions
Kolmogorov-Smirnov two-sample test	*D* = Largest difference between cumulative frequency distributions
Sign test	The difference in the number of cases in which the difference between the pairs in the sample is negative and the number of cases in which it is negative
Wilcoxon matched-pairs Rank Sum Test	W = Sum of negative signed ranks of $\lvert \text{rank - median} \rvert$
Wilcoxon 2-sample Rank Sum Test	W = Sum of ranks of smaller samples
Mann-Whitney *U* Test	*U* is the smaller of $U_1 = n_1 n_2 + \frac{n_1(n_1+1)}{2} - R_1$ $U_2 = n_1 n_2 + \frac{n_2(n_2+1)}{2} - R_2$

Kruskal-Wallis one-way analysis of variance	$H = \frac{12}{N(N+1)}\left(\sum \frac{R_j^2}{N_j}\right) - 3(N+1)$
Friedman's two-way analysis of variance by rank	$M = \frac{12}{NK(K+1)}\sum_{i=1}^{K} R_i^2 - 3N(K+1)$

Analysis of Variance

$$SS_R = \sum_{j=1}^{n_1}\left(x_{1j} - \overline{x}_1\right)^2 + \sum_{j=1}^{n_2}\left(x_{2j} - \overline{x}_2\right)^2 + \ldots + \sum_{j=1}^{n_k}\left(x_{kj} - \overline{x}_k\right)^2 = \sum_{i=1}^{k}\sum_{j=1}^{n_i}\left(x_{ij} - \overline{x}_i\right)^2$$

$$MS_R = \frac{SS_R}{n-k}$$

$$SS_B = n_1(\overline{x}_1 - \overline{x})^2 + n_2(\overline{x}_2 - \overline{x})^2 + \ \ldots \ + n_k(\overline{x}_k - \overline{x})^2$$

$$MS_B = \frac{SS_B}{k-1}$$

Test statistic $F = \frac{MS_B}{MS_W}$, with $k - 1$, $n - k$ degrees of freedom

Probability generating functions

$$G(t) = E(t^X) = \sum_r t^r P(X = r).$$

$$G(1) = 1$$

$$E(X) = \mu = G'(1)$$

$$Var(X) = G''(1) + \mu - \mu^2$$

Moment generating functions

$M_X(t) = E\left(e^{Xt}\right) = \sum_i e^{x_i t} p(x_i)$ (for a discrete variable)

$= \int_{\text{all } x} e^{xt} f(x) dx$ (for a continuous variable).

$$M_{A+B}(t) = M_A(t) \times M_B(t)$$

Index numbers

$$\textit{Laspeyre's Index} = \frac{\sum p_n q_0}{\sum p_0 q_0}$$

$$\textit{Paasche's Index} = \frac{\sum p_n q_n}{\sum p_0 q_n}$$

$$\textit{Typical Year Index} = \frac{\sum p_n q_t}{\sum p_0 q_t}$$

$$\textit{Fisher's Ideal Index} = \sqrt{\left\{\frac{\sum p_n q_0}{\sum p_0 q_0}\right\}\left\{\frac{\sum p_n q_n}{\sum p_0 q_n}\right\}}$$

where p_o = base year price, p_n = current year price,
q_o = base year weighting q_n = current year weighting
q_t = typical year weighting

Tables

Table 1. Normal distribution tables

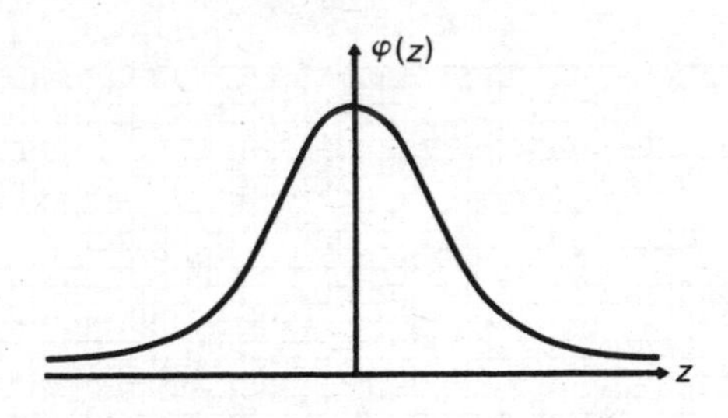

a. Table of $\varphi(\chi)$

χ	$\varphi(\chi)$	χ	$\varphi(\chi)$	χ	$\varphi(\chi)$	χ	$\varphi(\chi)$
0.0	.399	0.8	.290	1.6	.111	2.4	.022
0.1	.397	0.9	.266	1.7	.094	2.5	.018
0.2	.391	1.0	.242	1.8	.079	2.6	.014
0.3	.381	1.1	.218	1.9	.066	2.7	.010
0.4	.368	1.2	.194	2.0	.054	2.8	.008
0.5	.352	1.3	.171	2.1	.044	2.9	.006
0.6	.333	1.4	.150	2.2	.035	3.0	.004
0.7	.312	1.5	.130	2.3	.028		

Table 2(a). The c.d.f. of the standard normal distribution

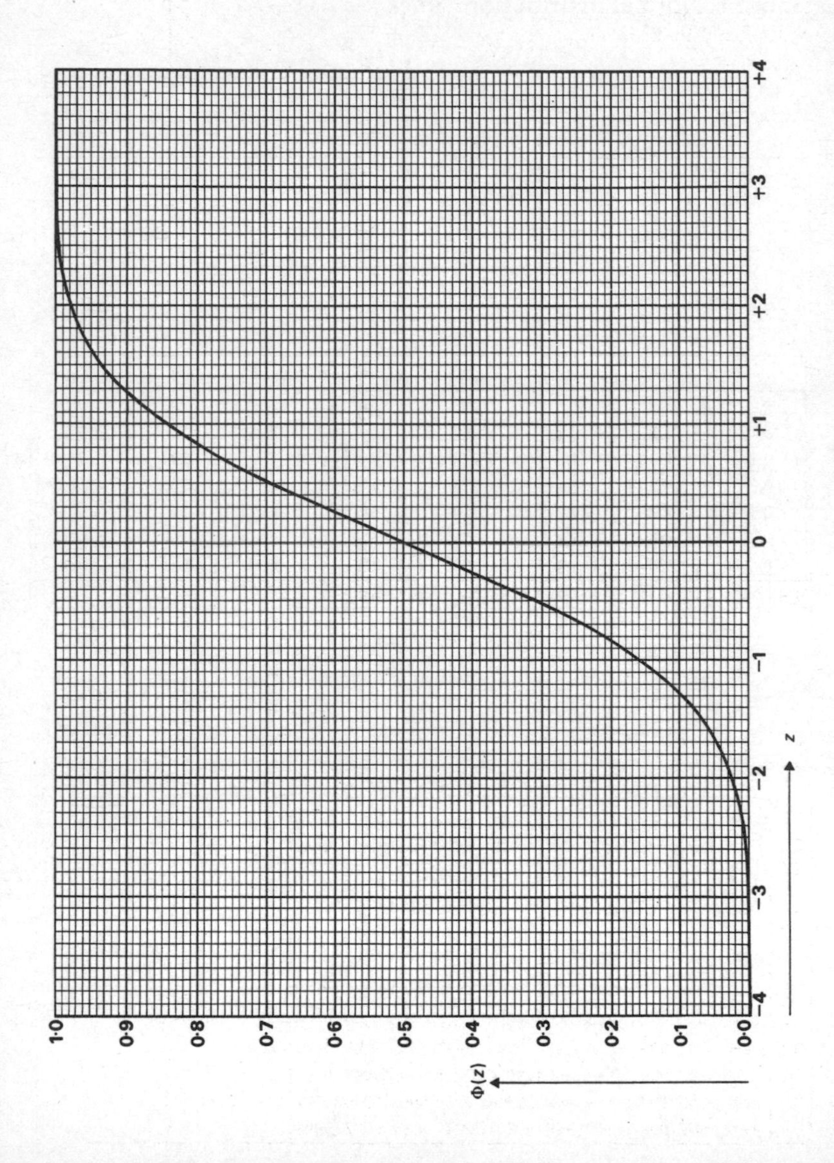

Table 2(b). The Normal distribution: values of $\Phi(z) = p$.

The table gives the probability, p, of a random variable distributed as N(0, 1) being less that z.

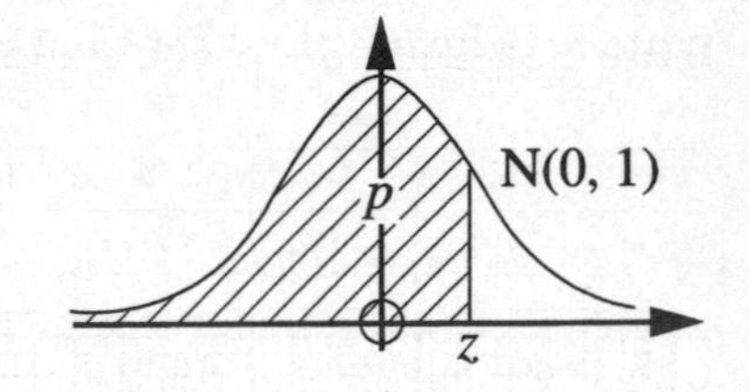

z	.00	.01	.02	.03	.04	.05	.06	.07	.08	.09	1	2	3	4	5	6	7	8	9
0.0	.5000	5040	5080	5120	5160	5199	5239	5279	5319	5359	4	8	12	16	20	24	28	32	36
0.1	.5398	5438	5478	5517	5557	5596	5636	5675	5714	5753	4	8	12	16	20	24	28	32	35
0.2	.5793	5832	5871	5910	5948	5987	6026	6064	6103	6141	4	8	12	15	19	23	27	31	35
0.3	.6179	6217	6255	6293	6331	6368	6406	6443	6480	6517	4	8	11	15	19	23	26	30	34
0.4	.6554	6591	6628	6664	6700	6736	6772	6808	6844	6879	4	7	11	14	18	22	25	29	32
0.5	.6915	6950	6985	7019	7054	7088	7123	7157	7190	7224	3	7	10	14	17	21	24	27	31
0.6	.7257	7291	7324	7357	7389	7422	7454	7486	7517	7549	3	6	10	13	16	19	23	26	29
0.7	.7580	7611	7642	7673	7704	7734	7764	7794	7823	7852	3	6	9	12	15	18	21	24	27
0.8	.7881	7910	7939	7967	7995	8023	8051	8078	8106	8133	3	6	8	11	14	17	19	22	25
0.9	.8159	8186	8212	8238	8264	8289	8315	8340	8365	8369	3	5	8	10	13	15	18	20	23
1.0	.8413	8438	8461	8485	8508	8531	8554	8577	8599	8621	2	5	7	9	12	14	16	18	21
1.1	.8643	8665	8686	8708	8729	8749	8770	8790	8810	8830	2	4	6	8	10	12	14	16	19
1.2	.8849	8869	8888	8907	8925	8944	8962	8980	8997	9015	2	4	6	7	9	11	13	15	16
1.3	.9032	9049	9066	9082	9099	9115	9131	9147	9162	9177	2	3	5	6	8	10	11	13	14
1.4	.9192	9207	9222	9236	9251	9265	9279	9292	9306	9319	1	3	4	6	7	8	10	11	13
1.5	.9332	9345	9357	9370	9382	9394	9406	9418	9429	9441	1	2	4	5	6	7	8	10	11
1.6	.9452	9463	9474	9484	9495	9505	9515	9525	9535	9545	1	2	3	4	5	6	7	8	9
1.7	.9554	9564	9573	9582	9591	9599	9608	9616	9625	9633	1	2	3	3	4	5	6	7	8
1.8	.9641	9649	9659	9664	9671	9678	9686	9693	9699	9706	1	1	2	3	4	4	5	6	6
1.9	.9713	9719	9726	9732	9738	9744	9750	9756	9761	9767	1	1	2	2	3	4	4	5	5
2.0	.9772	9778	9783	9788	9793	9798	9803	9808	9812	9817	0	1	1	2	2	3	3	4	4
2.1	.9821	9826	9830	9834	9838	9842	9846	9850	9854	9857	0	1	1	2	2	2	3	3	4
2.2	.9861	9864	9868	9871	9875	9878	9881	9884	9887	9890	0	1	1	1	2	2	2	3	3
2.3	.9893	9896	9898	9901	9904	9906	9909	9911	9913	9916	0	1	1	1	1	2	2	2	2
2.4	.9981	9920	9922	9925	9927	9929	9931	9932	9934	9936	0	0	1	1	1	1	1	2	2
2.5	.9938	9940	9941	9943	9945	9946	9948	9949	9951	9952									
2.6	.9953	9955	9956	9957	9959	9960	9961	9962	9963	9964									
2.7	.9965	9966	9967	9968	9969	9970	9971	9972	9973	9974									
2.8	.9974	9975	9976	9977	9977	9978	9979	9979	9980	9981									
2.9	.9981	9982	9982	9983	9984	9984	9985	9985	9986	9986									
3.0	.9987	9987	9989	9988	9988	9989	9989	9989	9990	9990									
3.1	.9990	9991	9991	9991	9992	9992	9992	9992	9993	9993									
3.2	.9993	9993	9994	9994	9994	9994	9994	9995	9995	9995									
3.3	.9995	9995	9996	9996	9996	9996	9996	9996	9996	9997									
3.4	.9997	9997	9997	9997	9997	9997	9997	9997	9997	9998									

differences untrustworthy

Table 3. Critical values for the *t* test

1-tail	10%	5%	2.5%	1%	0.5%	0.05%
2-tail	20%	10%	5%	2%	1%	0.1%
v						
1	3.078	6.314	12.706	31.821	63.657	636.619
2	1.886	2.920	4.303	6.965	9.925	31.598
3	1.638	2.353	3.182	4.541	5.841	12.941
4	1.533	2.132	2.776	3.747	4.604	8.610
5	1.476	2.015	2.571	3.365	4.032	6.859
6	1.440	1.943	2.447	3.143	3.707	5.959
7	1.415	1.895	2.365	2.998	3.499	5.405
8	1.397	1.860	2.306	2.896	3.355	5.041
9	1.383	1.833	2.262	2.821	3.250	4.781
10	1.372	1.812	2.228	2.764	3.169	4.587
11	1.363	1.796	2.201	2.718	3.106	4.437
12	1.356	1.782	2.179	2.681	3.055	4.318
13	1.350	1.771	2.160	2.650	3.012	4.221
14	1.345	1.761	2.145	2.624	2.977	4.140
15	1.341	1.753	2.131	2.602	2.947	4.073
16	1.337	1.746	2.120	2.583	2.921	4.015
17	1.333	1.740	2.110	2.567	2.898	3.965
18	1.330	1.734	2.101	2.552	2.878	3.922
19	1.328	1.729	2.093	2.539	2.861	3.883
20	1.325	1.725	2.086	2.528	2.845	3.850
21	1.323	1.721	2.080	2.518	2.831	3.819
22	1.321	1.717	2.074	2.508	2.819	3.792
23	1.319	1.714	2.069	2.500	2.807	3.767
24	1.318	1.711	2.064	2.492	2.797	3.745
25	1.316	1.708	2.060	2.485	2.787	3.725
26	1.315	1.706	2.056	2.479	2.779	3.707
27	1.314	1.703	2.052	2.473	2.771	3.690
28	1.313	1.701	2.048	2.467	2.763	3.674
29	1.311	1.699	2.045	2.462	2.756	3.659
30	1.310	1.697	2.042	2.457	2.750	3.646
40	1.303	1.684	2.021	2.423	2.704	3.551
60	1.296	1.671	2.000	2.390	2.460	3.460
120	1.289	1.658	1.980	2.358	2.617	3.373
–	1.282	1.645	1.960	2.326	2.576	3.291

Critical values for the Normal distribution

Source: *Statistical Tables*, 6th edn., R. A. Fisher and F. Yeates, 1974, Longman.

Table 4. Percentage points of the χ^2 (chi-squared) distribution

$p\%$	99	97.5	95	90	10	5.0	2.5	1.0	0.5
$v = 1$	.0001	.0010	.0039	.0158	2.706	3.841	5.024	6.635	7.879
2	.0201	.0506	0.103	0.211	4.605	5.991	7.378	9.210	10.60
3	0.115	0.216	0.352	0.584	6.251	7.815	9.348	11.34	12.84
4	0.297	0.484	0.711	1.064	7.779	9.488	11.14	13.28	14.86
5	0.554	0.831	1.145	1.610	9.236	11.07	12.83	15.09	16.75
6	0.872	1.237	1.635	2.204	10.64	12.59	14.45	16.81	18.55
7	1.239	1.690	2.167	2.833	12.02	14.07	16.01	18.48	20.28
8	1.646	2.180	2.733	3.490	13.36	15.51	17.53	20.09	21.95
9	2.088	2.700	3.325	4.168	14.68	16.92	19.02	21.67	23.59
10	2.558	3.247	3.940	4.865	15.99	18.31	20.48	23.21	25.19
11	3.053	3.816	4.575	5.578	17.28	19.68	21.92	24.72	26.76
12	3.571	4.404	5.226	6.304	18.55	21.03	23.34	26.22	28.30
13	4.107	5.009	5.892	7.042	19.81	22.36	24.74	27.69	29.82
14	4.660	5.629	6.571	7.790	21.06	23.68	26.12	29.14	31.32
15	5.229	6.262	7.261	8.547	22.31	25.00	27.49	30.58	32.80
16	5.812	6.908	7.962	9.312	23.54	26.30	28.85	32.00	34.27
17	6.408	7.564	8.672	10.09	24.77	27.59	30.19	33.41	35.72
18	7.015	8.231	9.390	10.86	25.99	28.87	31.53	34.81	37.16
19	7.633	8.907	10.12	11.65	27.20	30.14	32.85	36.19	38.58
20	8.260	9.591	10.85	12.44	28.41	31.41	34.17	37.57	40.00
21	8.897	10.28	11.59	13.24	29.62	32.67	35.48	38.93	41.40
22	9.542	10.98	12.34	14.04	30.81	33.92	36.78	40.29	42.80
23	10.20	11.69	13.09	14.85	32.01	35.17	38.08	41.64	44.18
24	10.86	12.40	13.85	15.66	33.20	36.42	39.36	42.98	45.56
25	11.52	13.12	14.61	16.47	34.38	37.65	40.65	44.31	46.93
26	12.20	13.84	15.38	17.29	35.56	38.89	41.92	45.64	48.29
27	12.88	14.57	16.15	18.11	36.74	40.11	43.19	46.96	49.64
28	13.56	15.31	16.93	18.94	37.92	41.34	44.46	48.28	50.99
29	14.26	16.05	17.71	19.77	39.09	42.56	45.72	49.59	52.34
30	14.95	16.79	18.49	20.60	40.26	43.77	46.98	50.89	53.67
35	18.51	20.57	22.47	24.80	46.06	49.80	53.20	57.34	60.27
40	22.16	24.43	26.51	29.05	51.81	55.76	59.34	63.69	66.77
50	29.71	32.36	34.76	37.69	63.17	67.50	71.42	76.15	79.49
100	70.06	74.22	77.93	82.36	118.5	124.3	129.6	135.8	140.2

Table 5(a). 5% points of the *F*-distribution

v_2 \ v_1	1	2	3	4	5	6	7	8	10	12	24	∞
1	161.4	199.5	215.7	224.6	230.2	234.0	236.8	238.9	241.9	243.9	249.0	254.3
2	18.5	19.0	19.2	19.2	19.3	19.3	19.4	19.4	19.4	19.4	19.5	19.5
3	10.13	9.55	9.28	9.12	9.01	8.94	8.89	8.85	8.79	8.74	8.64	8.53
4	7.71	6.94	6.59	6.39	6.26	6.16	6.09	6.04	5.96	5.91	5.77	5.63
5	6.61	5.79	5.41	5.19	5.05	4.95	4.88	4.82	4.74	4.68	4.53	4.36
6	5.99	5.14	4.76	4.53	4.39	4.28	4.21	4.15	4.06	4.00	3.84	3.67
7	5.59	4.74	4.35	4.12	3.97	3.87	3.79	3.73	3.64	3.57	3.41	3.23
8	5.32	4.46	4.07	3.84	3.69	3.58	3.50	3.44	3.35	3.28	3.12	2.93
9	5.12	4.26	3.86	3.63	3.48	3.37	3.29	3.23	3.14	3.07	2.90	2.71
10	4.96	4.10	3.71	3.48	3.33	3.22	3.14	3.07	2.98	2.91	2.74	2.54
11	4.84	3.98	3.59	3.36	3.20	3.09	3.01	2.95	2.85	2.79	2.61	2.40
12	4.75	3.89	3.49	3.26	3.11	3.00	2.91	2.85	2.75	2.69	2.51	2.30
13	4.67	3.81	3.41	3.18	3.03	2.92	2.83	2.77	2.67	2.60	2.42	2.21
14	4.60	3.74	3.34	3.11	2.96	2.85	2.76	2.70	2.60	2.53	2.35	2.13
15	4.54	3.68	3.29	3.06	2.90	2.79	2.71	2.64	2.54	2.48	2.29	2.07
16	4.49	3.63	3.24	3.01	2.85	2.74	2.66	2.59	2.49	2.42	2.24	2.01
17	4.45	3.59	3.20	2.96	2.81	2.70	2.61	2.55	2.45	2.38	2.19	1.96
18	4.41	3.55	3.16	2.93	2.77	2.66	2.58	2.51	2.41	2.34	2.15	1.92
19	4.38	3.52	3.13	2.90	2.74	2.63	2.54	2.48	2.38	2.31	2.11	1.88
20	4.35	3.39	3.10	2.87	2.71	2.60	2.51	2.45	2.35	2.28	2.08	1.84
21	4.32	3.47	3.07	2.84	2.68	2.57	2.49	2.42	2.32	2.25	2.05	1.81
22	4.30	3.44	3.05	2.82	2.66	2.55	2.46	2.40	2.30	2.23	2.03	1.78
23	4.28	3.42	3.03	2.80	2.64	2.53	2.44	2.37	2.27	2.20	2.00	1.76
24	4.26	3.40	3.01	2.78	2.62	2.51	2.42	2.36	2.25	2.18	1.98	1.73
25	4.24	3.39	2.99	2.76	2.60	2.49	2.40	2.34	2.24	2.16	1.96	1.71
26	4.23	3.37	2.98	2.74	2.59	2.47	2.39	2.32	2.22	2.15	1.95	1.69
27	4.21	3.35	2.96	2.73	2.57	2.46	2.37	2.31	2.20	2.13	1.93	1.67
28	4.20	3.34	2.95	2.71	2.56	2.45	2.36	2.29	2.19	2.12	1.91	1.65
29	4.18	3.33	2.93	2.70	2.55	2.43	2.35	2.28	2.18	2.10	1.90	1.64
30	4.17	3.32	2.92	2.69	2.53	2.42	2.33	2.27	2.16	2.09	1.89	1.62
32	4.15	3.29	2.90	2.67	2.51	2.40	2.31	2.24	2.14	2.07	1.86	1.59
34	4.13	3.28	2.88	2.65	2.49	2.38	2.29	2.23	2.12	2.05	1.84	1.57
36	4.11	3.26	2.87	2.63	2.48	2.36	2.28	2.21	2.11	2.03	1.82	1.55
38	4.10	3.24	2.85	2.62	2.46	2.35	2.26	2.19	2.09	2.02	1.81	1.53
40	4.08	3.23	2.84	2.61	2.45	2.34	2.25	2.18	2.08	2.00	1.79	1.51
60	4.00	3.15	2.76	2.53	2.37	2.25	2.17	2.10	1.99	1.92	1.70	1.39
120	3.92	3.07	2.68	2.45	2.29	2.18	2.09	2.02	1.91	1.83	1.61	1.25
∞	3.84	3.00	2.60	2.37	2.21	2.10	2.01	1.94	1.83	1.75	1.52	1.00

Source: *Cambridge Elementary Statistics Tables,* D.V. Lindley & J.C.P. Miller 1953 Cambridge University Press

Table 5(b). $2\frac{1}{2}$% points of the F-distribution

v_2 \ v_1	1	2	3	4	5	6	7	8	10	12	24	∞
1	648	800	864	900	922	937	948	957	969	977	997	1018
2	38.5	39.0	39.2	39.2	39.3	39.3	39.4	39.4	39.4	39.4	39.5	39.5
3	17.4	16.0	15.4	15.1	14.9	14.7	14.6	14.5	14.4	14.3	14.1	13.9
4	12.22	10.65	9.98	9.60	9.36	9.20	9.07	8.98	8.84	8.75	8.51	8.26
5	10.01	8.43	7.76	7.39	7.15	6.98	6.85	6.76	6.62	6.52	6.28	6.02
6	8.81	7.26	6.60	6.23	5.99	5.82	5.70	5.60	5.46	5.37	5.12	4.85
7	8.07	6.54	5.89	5.52	5.29	5.12	4.99	4.90	4.76	4.67	4.42	4.14
8	7.57	6.06	5.42	5.05	4.82	4.65	4.53	4.43	4.30	4.20	3.95	3.67
9	7.21	5.71	5.08	4.72	4.48	4.32	4.20	4.10	3.96	3.87	3.61	3.33
10	6.94	5.46	4.83	4.47	4.24	4.07	3.95	3.85	3.72	3.62	3.37	3.08
11	6.72	5.26	4.63	4.28	4.04	3.88	3.76	3.66	3.53	3.43	3.17	2.88
12	6.55	5.10	4.47	4.12	3.89	3.73	3.61	3.51	3.37	3.28	3.02	2.72
13	6.41	4.97	4.35	4.00	3.77	3.60	3.48	3.39	3.25	3.15	2.89	2.60
14	6.30	4.86	4.24	3.89	3.66	3.50	3.38	3.29	3.15	3.05	2.79	2.49
15	6.20	4.76	4.15	3.80	3.58	3.41	3.29	3.20	3.06	2.96	2.70	2.40
16	6.12	4.69	4.08	3.73	3.50	3.34	3.22	3.12	2.99	2.89	2.63	2.32
17	6.04	4.62	4.01	3.66	3.44	3.28	3.16	3.06	2.92	2.82	2.56	2.25
18	5.98	4.56	3.95	3.61	3.38	3.22	3.10	3.01	2.87	2.77	2.50	2.19
19	5.92	4.51	3.90	3.56	3.33	3.17	3.05	2.96	2.82	2.72	2.45	2.13
20	5.87	4.46	3.86	3.51	3.29	3.13	3.01	2.91	2.77	2.68	2.41	2.09
21	5.83	4.42	3.82	3.48	3.25	3.09	2.97	2.87	2.73	2.64	2.37	2.04
22	5.79	4.38	3.78	3.44	3.22	3.05	2.93	2.84	2.70	2.60	2.33	2.00
23	5.75	4.35	3.75	3.41	3.18	3.02	2.90	2.81	2.67	2.57	2.30	1.97
24	5.72	4.32	3.72	3.38	3.15	2.99	2.87	2.78	2.64	2.54	2.27	1.94
25	5.69	4.29	3.69	3.35	3.13	2.97	2.85	2.75	2.61	2.51	2.24	1.91
26	5.66	4.27	3.67	3.33	3.10	2.94	2.82	2.73	2.59	2.49	2.22	1.88
27	5.63	4.42	3.65	3.31	3.08	2.92	2.80	2.71	2.57	2.47	2.19	1.85
28	5.61	4.22	3.63	3.29	3.06	2.90	2.78	2.69	2.55	2.45	2.17	1.83
29	5.59	4.20	3.61	3.27	3.04	2.88	2.76	2.67	2.35	2.43	2.15	1.81
30	5.57	4.18	3.59	3.25	3.03	2.87	2.75	2.65	2.51	2.41	2.14	1.79
32	5.53	4.15	3.56	3.22	3.00	2.84	2.72	2.62	2.48	2.38	2.10	1.75
34	5.50	4.12	3.53	3.19	2.97	2.81	2.69	2.59	2.45	2.35	2.08	1.72
36	5.47	4.09	3.51	3.17	2.94	2.79	2.66	2.57	2.43	2.33	2.05	1.69
38	5.45	4.07	3.48	3.15	2.92	2.76	2.64	2.55	2.41	2.31	2.03	1.66
40	5.42	4.05	3.46	3.13	2.90	2.74	2.62	2.53	2.39	2.29	2.01	1.64
60	5.29	3.93	3.34	3.01	2.79	2.63	2.51	2.41	2.27	2.17	1.88	1.48
120	5.15	3.80	3.23	2.89	2.67	2.52	2.39	2.30	2.16	2.05	1.76	1.31
∞	5.02	3.69	3.12	2.79	2.57	2.41	2.29	2.19	2.05	1.94	1.64	1.00

Table 5(c). 1% points of the F-distribution

v_2 \ v_1	1	2	3	4	5	6	7	8	10	12	24	∞
1	4052	5000	5403	5625	5764	5859	5928	5981	6056	6106	6235	6366
2	98.5	99.0	99.2	99.2	99.3	99.3	99.4	99.4	99.4	99.4	99.5	99.5
3	34.1	30.8	29.5	28.7	28.2	27.9	27.7	27.5	27.2	27.1	26.6	26.1
4	21.2	18.0	16.7	16.0	15.5	15.2	15.0	14.8	14.5	14.4	13.9	13.5
5	16.26	13.27	12.06	11.39	10.97	10.67	10.46	10.29	10.05	9.89	9.47	9.02
6	13.74	10.92	9.78	9.15	8.75	8.47	8.26	8.10	7.87	7.72	7.31	6.88
7	12.25	9.55	8.45	7.85	7.46	7.19	6.99	6.84	6.62	6.47	6.07	5.65
8	11.26	8.65	7.59	7.01	6.63	6.37	6.18	6.03	5.81	5.67	5.28	4.86
9	10.56	8.02	6.99	6.42	6.06	5.80	5.61	5.47	5.26	5.11	4.73	4.31
10	10.04	7.56	6.55	5.99	5.64	5.39	5.20	5.06	4.85	4.71	4.33	3.91
11	9.65	7.21	6.22	5.67	5.32	5.07	4.89	4.74	4.54	4.40	4.02	3.60
12	9.33	6.93	5.95	5.41	5.06	4.82	4.64	4.50	4.30	4.16	3.78	3.36
13	9.07	6.70	5.74	5.21	4.86	4.62	4.44	4.30	4.10	3.96	3.59	3.17
14	8.86	6.51	5.56	5.04	4.70	4.46	4.28	4.14	3.94	3.80	3.43	3.00
15	8.68	6.36	5.42	4.89	4.56	4.32	4.14	4.00	3.80	3.67	3.29	2.87
16	8.53	6.23	5.29	4.77	4.44	4.20	4.03	3.89	3.69	3.55	3.18	2.75
17	8.40	6.11	5.18	4.67	4.34	4.10	3.93	3.79	3.59	3.46	3.08	2.65
18	8.29	6.01	5.09	4.58	4.25	4.01	3.84	3.71	3.51	3.37	3.00	2.57
19	8.18	5.93	5.01	4.50	4.17	3.94	3.77	3.63	3.43	3.30	2.92	2.49
20	8.10	5.85	4.94	4.43	4.10	3.87	3.70	3.56	3.37	3.23	2.86	2.42
21	8.02	5.78	4.87	4.37	4.04	3.81	3.64	3.51	3.31	3.17	2.80	2.36
22	7.95	5.72	4.82	4.31	3.99	3.76	3.59	3.45	3.26	3.12	2.75	2.31
23	7.88	5.66	4.76	4.26	3.94	3.71	3.54	3.41	3.21	3.07	2.70	2.26
24	7.82	5.61	4.72	4.22	3.90	3.67	3.50	3.36	3.17	3.03	2.66	2.21
25	7.77	5.57	4.68	4.18	3.86	3.63	3.46	3.32	3.13	2.99	2.62	2.17
26	7.72	5.53	4.64	4.14	3.82	3.59	3.42	3.29	3.09	2.96	2.58	2.13
27	7.68	5.49	4.60	4.11	3.78	3.56	3.39	3.26	3.06	2.93	2.55	2.10
28	7.64	5.45	4.57	4.07	3.75	3.53	3.36	3.23	3.03	2.90	2.52	2.06
29	7.60	5.42	4.54	4.04	3.73	3.50	3.33	3.20	3.00	2.87	2.49	2.03
30	7.56	5.39	4.51	4.02	3.70	3.47	3.30	3.17	2.98	2.84	2.47	2.01
32	7.50	5.34	4.46	3.97	3.65	3.43	3.26	3.13	2.93	2.80	2.42	1.96
34	7.45	5.29	4.42	3.93	3.61	3.39	3.22	3.09	2.90	2.76	2.38	1.91
36	7.40	5.25	4.38	3.89	3.58	3.35	3.18	3.05	2.86	2.72	2.35	1.87
38	7.35	5.21	4.34	3.86	3.54	3.32	3.15	3.02	2.83	2.69	2.32	1.84
40	7.31	5.18	4.31	3.83	3.51	3.29	3.12	2.99	2.80	2.66	2.29	1.80
60	7.08	4.98	4.13	3.65	3.34	3.12	2.95	2.82	2.63	2.50	2.12	1.60
120	6.85	4.79	3.95	3.48	3.17	2.96	2.79	2.66	2.47	2.34	1.95	1.38
∞	6.63	4.61	3.78	3.32	3.02	2.80	2.64	2.51	2.32	2.18	1.79	1.00

Table 5(d). 0.1% points of the F-distribution

v_2 \ v_1	1	2	3	4	5	6	7	8	10	12	24	∞
1	4053	5000	5404	5625	5764	5859	5929	5981	6056	6107	6235	6366
2	998.5	999.0	999.2	999.2	999.3	999.3	999.4	999.4	999.4	999.4	999.5	999.5
3	167.0	148.5	141.1	137.1	134.6	132.8	131.5	130.6	129.2	128.3	125.9	123.5
4	74.14	61.25	56.18	53.44	51.71	50.53	49.66	49.00	48.05	47.41	45.77	44.05
5	47.18	37.12	33.20	31.09	29.75	28.83	28.16	27.65	26.92	26.42	25.14	23.79
6	35.51	27.00	23.70	21.92	20.80	20.03	19.46	19.03	18.41	17.99	16.90	15.75
7	29.25	21.69	18.77	17.20	16.21	15.52	15.02	14.63	14.08	13.71	12.73	11.70
8	25.42	18.49	15.83	14.39	13.48	12.86	12.40	12.05	11.54	11.19	10.30	9.34
9	22.86	16.39	13.90	12.56	11.71	11.31	10.69	10.37	9.87	9.57	8.72	7.81
10	21.04	14.91	12.55	11.28	10.48	9.93	9.52	9.20	8.74	8.44	7.64	6.76
11	19.69	13.81	11.56	10.35	9.58	9.05	8.66	8.35	7.92	7.63	6.85	6.00
12	18.64	12.97	10.80	9.63	8.89	8.38	8.00	7.71	7.29	7.00	6.25	5.42
13	17.82	12.31	10.21	9.07	8.35	7.86	7.49	7.21	6.80	6.52	5.78	4.97
14	17.14	11.78	9.73	8.62	7.92	7.44	7.08	6.80	6.40	6.13	5.41	4.60
15	16.59	11.34	9.34	8.25	7.57	7.09	6.74	6.47	6.08	5.81	5.10	4.31
16	16.12	10.97	9.01	7.94	7.27	6.80	6.46	6.19	5.81	5.55	4.85	4.06
17	15.72	10.66	8.73	7.86	7.02	6.56	6.22	5.96	5.58	5.32	4.63	3.85
18	15.38	10.39	8.49	7.46	6.81	6.35	6.02	5.76	5.39	5.13	4.45	3.67
19	15.08	10.16	8.28	7.27	6.62	6.18	5.85	5.59	5.22	4.97	4.29	3.51
20	14.82	9.95	8.10	7.10	6.46	6.02	5.69	5.44	5.08	4.82	4.15	3.38
21	14.59	9.77	7.94	6.95	6.32	5.88	5.56	5.31	4.95	4.70	4.03	3.26
22	14.38	9.61	7.80	6.81	6.19	5.76	5.44	5.19	4.83	4.58	3.92	3.15
23	14.19	9.47	7.67	6.70	6.08	5.65	5.33	5.09	4.73	4.48	3.82	3.05
24	14.03	9.34	7.55	6.59	5.98	5.55	5.23	4.99	4.64	4.39	3.74	2.97
25	13.88	9.22	7.45	6.49	5.89	5.46	5.15	4.91	4.56	4.31	3.66	2.89
26	13.74	9.12	7.36	6.41	5.80	5.38	5.07	4.83	4.48	4.24	3.59	2.82
27	13.61	9.02	7.27	6.33	5.73	5.31	5.00	4.76	4.41	4.17	3.52	2.75
28	13.50	8.93	7.19	6.25	5.66	5.24	4.93	4.69	4.35	4.11	3.46	2.69
29	13.39	8.85	7.12	6.19	5.59	5.18	4.87	4.64	4.29	4.05	3.41	2.64
30	13.29	8.77	7.05	6.12	5.53	5.12	4.82	4.58	4.24	4.00	3.36	2.59
32	13.12	8.64	6.94	6.01	5.43	5.02	4.72	4.48	4.14	3.91	3.27	2.50
34	12.97	8.52	6.83	5.92	5.34	4.93	4.63	4.40	4.06	3.83	3.19	2.42
36	12.83	8.42	6.74	5.84	5.26	4.86	4.56	4.33	3.99	3.76	3.12	2.35
38	12.71	8.33	6.66	5.76	5.19	4.79	4.49	4.26	3.93	3.70	3.06	2.29
40	12.61	8.25	6.59	5.70	5.13	4.73	4.44	4.21	3.87	3.64	3.01	2.23
60	11.97	7.77	6.17	5.31	4.76	4.37	4.09	3.86	3.54	3.32	2.69	1.89
120	11.38	7.32	5.78	4.95	4.42	4.04	3.77	3.55	3.24	3.02	2.40	1.54
∞	10.83	6.91	5.42	4.62	4.10	3.74	3.47	3.27	2.96	2.74	2.13	1.00

Table 6. Critical values for Pearson's product moment correlation coefficient, *r*

1-tail test	5%	2½%	1%	½%
2-tail test	10%	5%	2%	1%
n				
1	–	–	–	–
2	–	–	–	–
3	0.9877	0.9969	0.9995	0.9999
4	0.9000	0.9500	0.9800	0.9900
5	0.8054	0.8783	0.9343	0.9587
6	0.7293	0.8114	0.8822	0.9172
7	0.6694	0.7545	0.8329	0.8745
8	0.6215	0.7067	0.7887	0.8343
9	0.5822	0.6664	0.7498	0.7977
10	0.5494	0.6319	0.7155	0.7646
11	0.5214	0.6021	0.6851	0.7348
12	0.4973	0.5760	0.6581	0.7079
13	0.4762	0.5529	0.6339	0.6835
14	0.4575	0.5324	0.6120	0.6614
15	0.4409	0.5140	0.5923	0.6411
16	0.4259	0.4973	0.5742	0.6226
17	0.4124	0.4821	0.5577	0.6055
18	0.4000	0.4683	0.5425	0.5897
19	0.3887	0.4555	0.5285	0.5751
20	0.3783	0.4438	0.5155	0.5614
21	0.3687	0.4329	0.5034	0.5487
22	0.3598	0.4227	0.4921	0.5368
23	0.3515	0.4132	0.4815	0.5256
24	0.3438	0.4044	0.4716	0.5151
25	0.3365	0.3961	0.4622	0.5052
26	0.3297	0.3882	0.4534	0.4958
27	0.3233	0.3809	0.4451	0.4869
28	0.3172	0.3739	0.4372	0.4785
29	0.3115	0.3673	0.4297	0.4705
30	0.3061	0.3610	0.4226	0.4629
31	0.3009	0.3550	0.4158	0.4556
32	0.2960	0.3494	0.4093	0.4487
33	0.2913	0.3440	0.4032	0.4421
34	0.2869	0.3388	0.3972	0.4357
35	0.2826	0.3388	0.3916	0.4926
36	0.2785	0.3291	0.3862	0.4238
37	0.2746	0.3246	0.3810	0.4182
38	0.2709	0.3202	0.3760	0.4128
39	0.2673	0.3160	0.3712	0.4076
40	0.2638	0.3120	0.3665	0.4026
41	0.2605	0.3081	0.3621	0.3978
42	0.2573	0.3044	0.3578	0.3932
43	0.2542	0.3008	0.3536	0.3887
44	0.2512	0.2973	0.3496	0.3843
45	0.2483	0.2940	0.3457	0.3801
46	0.2455	0.2907	0.3420	0.3761
47	0.2429	0.2876	0.3384	0.3721
48	0.2403	0.2845	0.3348	0.3683
49	0.2377	0.2816	0.3314	0.3646
50	0.2353	0.2787	0.3281	0.3610
51	0.2329	0.2759	0.3249	0.3575
52	0.2306	0.2732	0.3218	0.3542
53	0.2284	0.2706	0.3188	0.3509
54	0.2262	0.2681	0.3158	0.3477
55	0.2241	0.2656	0.3129	0.3445
56	0.2221	0.2632	0.3102	0.3415
57	0.2201	0.2609	0.3074	0.3385
58	0.2181	0.2586	0.3048	0.3357
59	0.2162	0.2564	0.3022	0.3328
60	0.2144	0.2542	0.2997	0.3301

Table 7. Critical values for Spearman's rank correlation coefficient, r_s

1-tail test	5%	2½%	1%	½%	1-tail test	5%	2½%	1%	½%
2-tail test	10%	5%	2%	1%	2-tail test	10%	5%	2%	1%
n					n				
1	–	–	–	–	31	0.3012	0.3560	0.4185	0.4593
2	–	–	–	–	32	0.2962	0.3504	0.4117	0.4523
3	–	–	–	–	33	0.2914	0.3449	0.4054	0.4455
4	1.0000	–	–	–	34	0.2871	0.3396	0.3995	0.4390
5	0.9000	1.0000	1.0000	–	35	0.2829	0.3347	0.3936	0.4328
6	0.8286	0.8857	0.9429	1.0000	36	0.2788	0.3300	0.3882	0.4268
7	0.7143	0.7857	0.8929	0.9286	37	0.2748	0.3253	0.3829	0.4211
8	0.6429	0.7381	0.8333	0.8810	38	0.2710	0.3209	0.3778	0.4155
9	0.6000	0.7000	0.7833	0.8333	39	0.2674	0.3168	0.3729	0.4103
10	0.5636	0.6485	0.7455	0.7939	40	0.2640	0.3128	0.3681	0.4051
11	0.5364	0.6182	0.7091	0.7545	41	0.2606	0.3087	0.3636	0.4002
12	0.5035	0.5874	0.6783	0.7273	42	0.2574	0.3051	0.3594	0.3955
13	0.4835	0.5604	0.6484	0.7033	43	0.2543	0.3014	0.3550	0.3908
14	0.4637	0.5385	0.6264	0.6791	44	0.2513	0.2978	0.3511	0.3865
15	0.4464	0.5214	0.6036	0.6536	45	0.2484	0.2945	0.3470	0.3882
16	0.4294	0.5029	0.5824	0.6353	46	0.2456	0.2913	0.3433	0.3781
17	0.4142	0.4877	0.5662	0.6176	47	0.2429	0.2880	0.3396	0.3741
18	0.4014	0.4716	0.5501	0.5996	48	0.2403	0.2850	0.3361	0.3702
19	0.3912	0.4596	0.5351	0.5842	49	0.2378	0.2820	0.3326	0.3664
20	0.3805	0.4466	0.5218	0.5699	50	0.2353	0.2791	0.3293	0.3628
21	0.3701	0.4364	0.5091	0.5558	51	0.2329	0.2764	0.3260	0.3592
22	0.3608	0.4252	0.4975	0.5438	52	0.2307	0.2736	0.3228	0.3558
23	0.3528	0.4160	0.4862	0.5316	53	0.2284	0.2710	0.3198	0.3524
24	0.3443	0.4070	0.4757	0.5209	54	0.2262	0.2685	0.3168	0.3492
25	0.3369	0.3977	0.4662	0.5108	55	0.2242	0.2659	0.3139	0.3460
26	0.3306	0.3901	0.4571	0.5009	56	0.2221	0.2636	0.3111	0.3429
27	0.3242	0.3828	0.4487	0.4915	57	0.2201	0.2612	0.3083	0.3400
28	0.3180	0.3755	0.4401	0.4828	58	0.2181	0.2589	0.3057	0.3370
29	0.3118	0.3685	0.4325	0.4749	59	0.2162	0.2567	0.3030	0.3342
30	0.3063	0.3624	0.4251	0.4670	60	0.2144	0.2545	0.3005	0.3314

Table 8. Critical values for Kendall's coefficient of rank correlation

1-tail test	5%	$2\frac{1}{2}$%	1%	$\frac{1}{2}$%
2-tail test	10%	5%	2%	1%
v				
1	–	–	–	–
2	1.0000	–	–	–
3	0.8000	1.0000	1.0000	–
4	0.7333	0.8667	0.8667	1.0000
5	0.6190	0.7143	0.8095	0.9048
6	0.5714	0.6429	0.7143	0.7857
7	0.5000	0.5556	0.6667	0.7222
8	0.4667	0.5111	0.6000	0.6444
9	0.4182	0.4909	0.5636	0.6000
10	0.3939	0.4545	0.5455	0.5758
11	0.3590	0.4359	0.5128	0.5641
12	0.3626	0.4066	0.4725	0.5165
13	0.3333	0.3905	0.4667	0.5048
14	0.3167	0.3833	0.4333	0.4833
15	0.3088	0.3676	0.4265	0.4706
16	0.2941	0.3464	0.4118	0.4510
17	0.2666	0.3333	0.3918	0.4386
18	0.2737	0.3263	0.3789	0.4211
19	0.2867	0.3143	0.3714	0.4095
20	0.2641	0.3074	0.3593	0.3939
21	0.2569	0.2964	0.3518	0.3913
22	0.2464	0.2899	0.3406	0.3768
23	0.2400	0.2867	0.3333	0.3667
24	0.2369	0.2800	0.3292	0.3600
25	0.2308	0.2707	0.3219	0.3561
26	0.2275	0.2646	0.3122	0.3439
27	0.2217	0.2611	0.3103	0.3399
28	0.2184	0.2552	0.3011	0.3333
29	0.2129	0.2516	0.2946	0.3247
30	0.2097	0.2460	0.2903	0.3226

Source: *Elementary Statistics Tables*, H.R. Neave 1981, George Allen & Unwin.

Table 9. Fisher's *z* transformation of *r* to *z*

r	z_r	r	z_r	r	z_r
.01	.010	.34	.354	.67	.811
.02	.020	.35	.366	.68	.829
.03	.030	.36	.377	.69	.848
.04	.040	.37	.389	.70	.867
.05	.050	.38	.400	.71	.887
.06	.060	.39	.412	.72	.908
.07	.070	.40	.424	.73	.929
.08	.080	.41	.436	.74	.950
.09	.090	.42	.448	.75	.973
.10	.100	.43	.460	.76	.996
.11	.110	.44	.472	.77	1.020
.12	.121	.45	.485	.78	1.045
.13	.131	.46	.497	.79	1.071
.14	.141	.47	.510	.80	1.099
.15	.151	.48	.523	.81	1.127
.16	.161	.49	.536	.82	1.157
.17	.172	.50	.549	.83	1.188
.18	.182	.51	.563	.84	1.221
.19	.192	.52	.577	.85	1.256
.20	.203	.53	.590	.86	1.293
.21	.214	.54	.604	.87	1.333
.22	.224	.55	.618	.88	1.376
.23	.234	.56	.633	.89	1.422
.24	.245	.57	.648	.90	1.472
.25	.256	.58	.663	.91	1.528
.26	.266	.59	.678	.92	1.589
.27	.277	.60	.693	.93	1.658
.28	.288	.61	.709	.94	1.738
.29	.299	.62	.725	.95	1.832
.30	.309	.63	.741	.96	1.946
.31	.321	.64	.758	.97	2.092
.32	.332	.65	.775	.98	2.298
.33	.343	.66	.793	.99	2.647

Source: *Psychology Statistics*, Q. McNemar, John Wiley & Sons, Inc.

Table 10. Conversion of range to standard deviation

n	a_n	n	a_n	n	a_n	n	a_n
2	0.8862	5	0.4299	8	0.3512	11	0.3152
3	0.5908	6	0.3946	9	0.3367	12	0.3069
4	0.4857	7	0.3698	10	0.3249	13	0.2998

Source: *Advanced General Statistics*, B.C. Erricker 1971, Hodder & Stoughton.

Table 11. Critical values for the Mann-Whitney U Test

1-tail		5%	2½%	1%	½%
2-tail		10%	5%	2%	1%
m	n				
2	2	–	–	–	–
2	3	–	–	–	–
2	4	–	–	–	–
2	5	0	–	–	–
2	6	0	–	–	–
2	7	0	–	–	–
2	8	1	0	–	–
2	9	1	0	–	–
2	10	1	0	–	–
2	11	1	0	–	–
2	12	2	1	–	–
2	13	2	1	0	–
2	14	3	1	0	–
2	15	3	1	0	–
2	16	3	1	0	–
2	17	3	2	0	–
2	18	4	2	0	–
2	19	4	2	1	0
2	20	4	2	1	0
2	21	5	3	1	0
2	22	5	3	1	0
2	23	5	3	1	0
2	24	6	3	1	0
2	25	6	3	1	0
3	3	0	–	–	–
3	4	0	–	–	–
3	5	1	0	–	–
3	6	2	1	–	–
3	7	2	1	0	–
3	8	3	2	0	–
3	9	4	2	1	0
3	10	4	3	1	0
3	11	5	3	1	0
3	12	5	4	2	1
3	13	6	4	2	1
3	14	7	5	2	1
3	15	7	5	3	2
3	16	8	6	3	2
3	17	9	6	4	2

1-tail		5%	2½%	1%	½%
2-tail		10%	5%	2%	1%
m	n				
3	18	9	7	4	2
3	19	10	7	4	3
3	20	11	8	5	3
3	21	11	8	5	3
3	22	12	9	6	4
3	23	13	9	6	4
3	24	13	10	6	4
3	25	14	10	7	5
4	4	1	0	–	–
4	5	2	1	0	–
4	6	3	2	1	0
4	7	4	3	1	0
4	8	5	4	2	1
4	9	6	4	3	1
4	10	7	5	3	2
4	11	8	6	4	2
4	12	9	7	5	3
4	13	10	8	5	3
4	14	11	9	6	4
4	15	12	10	7	5
4	16	14	11	7	5
4	17	15	11	8	6
4	18	16	12	9	6
4	19	17	13	9	7
4	20	18	14	10	8
4	21	19	15	11	8
4	22	20	16	11	9
4	23	21	17	12	9
4	24	22	17	13	10
4	25	23	18	13	10
5	5	4	2	1	0
5	6	5	3	2	1
5	7	6	5	3	1
5	8	8	6	4	2
5	9	9	7	5	3
5	10	11	8	6	4
5	11	12	9	7	5
5	12	13	11	8	6
5	13	15	12	9	7

Table 11. cont.

1-tail		5%	2½%	1%	½%
2-tail		10%	5%	2%	1%
m	*n*				
5	14	16	13	10	7
5	15	18	14	11	8
5	16	19	15	12	9
5	17	20	17	13	10
5	18	22	18	14	11
5	19	23	19	15	12
5	20	25	20	16	13
5	21	26	22	17	14
5	22	28	23	18	14
5	23	29	24	19	15
5	24	30	25	20	16
5	25	32	27	21	17
6	6	7	5	3	2
6	7	8	6	4	3
6	8	10	8	6	4
6	9	12	10	7	5
6	10	14	11	8	6
6	11	16	13	9	7
6	12	17	14	11	9
6	13	19	16	12	10
6	14	21	17	13	11
6	15	23	19	15	12
6	16	25	21	16	13
6	17	26	22	18	15
6	18	28	24	19	16
6	19	30	25	20	17
6	20	32	27	22	18
6	21	34	29	23	19
6	22	36	30	24	21
6	23	37	32	26	22
6	24	39	33	27	23
6	25	41	35	29	24
7	7	11	8	6	4
7	8	13	10	7	6
7	9	15	12	9	7
7	10	17	14	11	9
7	11	19	16	12	10
7	12	21	18	14	12
7	13	24	20	16	13
7	14	26	22	17	15
7	15	28	24	19	16
7	16	30	26	21	18
7	17	33	28	23	19
7	18	35	30	24	21
7	19	37	32	26	22
7	20	39	34	28	24
7	21	41	36	30	25
7	22	44	38	31	27
7	23	46	40	33	29
7	24	48	42	35	30
7	25	50	44	36	32
8	8	15	13	9	7
8	9	18	15	11	9
8	10	20	17	13	11
8	11	23	19	15	13
8	12	26	22	17	15
8	13	28	24	20	17
8	14	31	26	22	18
8	15	33	29	24	20
8	16	36	31	26	22
8	17	39	34	28	24
8	18	41	36	30	26
8	19	44	38	32	28
8	20	47	41	34	30
8	21	49	43	36	32
8	22	52	45	38	34
8	23	54	48	40	35
8	24	57	50	42	37
8	25	60	53	45	39
9	9	21	17	14	11
9	10	24	20	16	13
9	11	27	23	18	16
9	12	30	26	21	18
9	13	33	28	23	20
9	14	36	31	26	22
9	15	39	34	28	24
9	16	42	37	31	27
9	17	45	39	33	29

Table 11. cont.

1-tail		5%	2½%	1%	½%
2-tail		10%	5%	2%	1%
m	*n*				
9	18	48	42	35	31
9	19	51	45	38	33
9	20	54	48	40	36
9	21	57	50	43	38
9	22	60	53	45	40
9	23	63	56	48	43
9	24	66	59	50	45
9	25	69	62	53	47
10	10	27	23	19	16
10	11	31	26	22	18
10	12	34	29	24	21
10	13	37	33	27	24
10	14	41	36	30	26
10	15	44	39	33	29
10	16	48	42	36	31
10	17	51	45	38	34
10	18	55	48	41	37
10	19	58	52	44	39
10	20	62	55	47	42
10	21	65	58	50	44
10	22	68	61	53	47
10	23	72	64	55	50
10	24	75	67	58	52
10	25	79	71	61	55
11	11	34	30	25	21
11	12	38	33	28	24
11	13	42	37	31	27
11	14	46	40	34	30
11	15	50	44	37	33
11	16	54	47	41	36
11	17	57	51	44	39
11	18	61	55	47	42
11	19	65	58	50	45
11	20	69	62	53	48
11	21	73	65	57	51
11	22	77	69	60	54
11	23	81	73	63	57
11	24	85	76	66	60
11	25	89	80	70	63

1-tail		5%	2½%	1%	½%
2-tail		10%	5%	2%	1%
m	*n*				
12	12	42	37	31	27
12	13	47	41	35	31
12	14	51	45	38	34
12	15	55	49	42	37
12	16	60	53	46	41
12	17	64	57	49	44
12	18	68	61	53	47
12	19	72	65	56	51
12	20	77	69	60	54
12	21	81	73	64	58
12	22	85	77	67	61
12	23	90	81	71	64
12	24	94	85	75	68
12	25	98	89	78	71
13	13	51	45	39	34
13	14	56	50	43	38
13	15	61	54	47	42
13	16	65	59	51	45
13	17	70	63	55	49
13	18	75	67	59	53
13	19	80	72	63	57
13	20	84	76	67	60
13	21	89	80	71	64
13	22	94	85	75	68
13	23	98	89	79	72
13	24	103	94	83	75
13	25	108	98	87	79
14	14	61	55	47	42
14	15	66	59	51	46
14	16	71	64	56	50
14	17	77	69	60	54
14	18	82	74	65	58
14	19	87	78	69	63
14	20	92	83	73	67
14	21	97	88	78	71
14	22	102	93	82	75
14	23	107	98	87	79
14	24	113	102	97	83
14	25	118	107	95	87

Table 11. cont.

1-tail		5%	2½%	1%	½%
2-tail		10%	5%	2%	1%
m	n				
15	15	72	64	56	51
15	16	77	70	61	55
15	17	83	75	66	60
15	18	88	80	70	64
15	19	94	85	75	69
15	20	100	90	80	73
15	21	105	96	85	78
15	22	111	101	90	82
15	23	116	106	94	87
15	24	122	111	99	91
15	25	128	117	104	96
16	16	83	75	66	60
16	17	89	81	71	65
16	18	95	86	76	70
16	19	101	92	82	74
16	20	107	98	87	79
16	21	113	103	92	84
16	22	119	109	97	89
16	23	125	115	102	94
16	24	131	120	108	99
16	25	137	126	113	104
17	17	96	87	77	70
17	18	102	93	82	75
17	19	109	99	88	81
17	20	115	105	93	86
17	21	121	111	99	91
17	22	128	117	105	96
17	23	134	123	110	102
17	24	141	129	116	107
17	25	147	135	122	112
18	18	109	99	88	81
18	19	116	106	94	87
18	20	123	112	100	92
18	21	130	119	106	98

1-tail		5%	2½%	1%	½%
2-tail		10%	5%	2%	1%
m	n				
18	22	136	125	112	104
18	23	143	132	118	109
18	24	150	138	124	115
18	25	157	145	130	121
19	19	123	113	101	93
19	20	130	119	107	99
19	21	138	126	113	105
19	22	145	133	120	111
19	23	152	140	126	117
19	24	160	147	133	123
19	25	167	154	139	129
20	20	138	127	114	105
20	21	146	134	121	112
20	22	154	141	127	118
20	23	161	149	134	125
20	24	169	156	141	131
20	25	177	163	148	138
21	21	154	142	128	118
21	22	162	150	135	125
21	23	170	157	142	132
21	24	179	165	150	139
21	25	187	173	157	146
22	22	171	158	143	133
22	23	179	166	150	140
22	24	188	174	158	147
22	25	197	182	166	155
23	23	189	175	158	148
23	24	198	183	167	155
23	25	207	192	175	163
24	24	207	192	175	164
24	25	217	201	184	172
25	25	227	211	192	180

For larger values of m, n it is usually adequate to use a Normal approximation with continuity correction, with mean $\frac{1}{2}mn$ and variance $\frac{1}{12}mn(m+n+1)$.

Table 12. Critical values for the Wilcoxon Rank Sum 2-Sample Test

1-tail		5%	2½%	1%	½%
2-tail		10%	5%	2%	1%
m	*n*				
2	2	–	–	–	–
2	3	–	–	–	–
2	4	–	–	–	–
2	5	3	–	–	–
2	6	3	–	–	–
2	7	3	–	–	–
2	8	4	3	–	–
2	9	4	3	–	–
2	10	4	3	–	–
2	11	4	3	–	–
2	12	5	4	–	–
2	13	5	4	3	–
2	14	6	4	3	–
2	15	6	4	3	–
2	16	6	4	3	–
2	17	6	5	3	–
2	18	7	5	3	–
2	19	7	5	4	3
2	20	7	5	4	3
2	21	8	6	4	3
2	22	8	6	4	3
2	23	8	6	4	3
2	24	9	6	4	3
2	25	9	6	4	3
3	3	6	–	–	–
3	4	6	–	–	–
3	5	7	6	–	–
3	6	8	7	–	–
3	7	8	7	6	–
3	8	9	8	6	–
3	9	10	8	7	6
3	10	10	9	7	6
3	11	11	9	7	6
3	12	11	10	8	7
3	13	12	10	8	7
3	14	13	11	8	7
3	15	13	11	9	8
3	16	14	12	9	8
3	17	15	12	10	8

1-tail		5%	2½%	1%	½%
2-tail		10%	5%	2%	1%
m	*n*				
3	18	15	13	10	8
3	19	16	13	10	9
3	20	17	14	11	9
3	21	17	14	11	9
3	22	18	15	12	10
3	23	19	15	12	10
3	24	19	16	12	10
3	25	20	16	13	11
4	4	11	10	–	–
4	5	12	11	10	–
4	6	13	12	11	10
4	7	14	13	11	10
4	8	15	14	12	11
4	9	16	14	13	11
4	10	17	15	13	12
4	11	18	16	14	12
4	12	19	17	15	13
4	13	20	18	15	13
4	14	21	19	16	14
4	15	22	20	17	15
4	16	24	21	17	15
4	17	25	21	18	16
4	18	26	22	19	16
4	19	27	23	19	17
4	20	28	24	20	18
4	21	29	25	21	18
4	22	30	26	21	19
4	23	31	27	22	19
4	24	32	27	23	20
4	25	33	28	23	20
5	5	19	17	16	15
5	6	20	18	17	16
5	7	21	20	18	16
5	8	23	21	19	17
5	9	24	22	20	18
5	10	26	23	21	19
5	11	27	24	22	20
5	12	28	26	23	21

Table 12. contd.

1-tail		5%	2½%	1%	½%
2-tail		10%	5%	2%	1%
m	*n*				
5	13	30	27	24	22
5	14	31	28	25	22
5	15	33	29	26	23
5	16	34	30	27	24
5	17	35	32	28	25
5	18	37	33	29	26
5	19	38	34	30	27
5	20	40	35	31	28
5	21	41	37	32	29
5	22	43	38	33	29
5	23	44	39	34	30
5	24	45	40	35	31
5	25	47	42	36	32
6	6	28	26	24	23
6	7	29	27	25	24
6	8	31	29	27	25
6	9	33	31	28	26
6	10	35	32	29	27
6	11	37	34	30	28
6	12	38	35	32	30
6	13	40	37	33	31
6	14	42	38	34	32
6	15	44	40	36	33
6	16	46	42	37	34
6	17	47	43	39	36
6	18	49	45	40	37
6	19	51	46	41	38
6	20	53	48	43	39
6	21	55	50	44	40
6	22	57	51	45	42
6	23	58	53	47	43
6	24	60	54	48	44
6	25	62	56	50	45
7	7	39	36	34	32
7	8	41	38	35	34
7	9	43	40	37	35
7	10	45	42	39	37
7	11	47	44	40	38
7	12	49	46	42	40

1-tail		5%	2½%	1%	½%
2-tail		10%	5%	2%	1%
m	*n*				
7	13	52	48	44	41
7	14	54	50	45	43
7	15	56	52	47	44
7	16	58	54	49	46
7	17	61	56	51	47
7	18	63	58	52	49
7	19	65	60	54	50
7	20	67	62	56	52
7	21	69	64	58	53
7	22	72	66	59	55
7	23	74	68	61	57
7	24	76	70	63	58
7	25	78	72	64	60
8	8	51	49	45	43
8	9	54	51	47	45
8	10	56	53	49	47
8	11	59	55	51	49
8	12	62	58	53	51
8	13	64	60	56	53
8	14	67	62	58	54
8	15	69	65	60	56
8	16	72	67	62	58
8	17	75	70	64	60
8	18	77	72	66	62
8	19	80	74	68	64
8	20	83	77	70	66
8	21	85	79	72	68
8	22	88	81	74	70
8	23	90	84	76	71
8	24	93	86	78	73
8	25	96	89	81	75
9	9	66	62	59	56
9	10	69	65	61	58
9	11	72	68	63	61
9	12	75	71	66	63
9	13	78	73	68	65
9	14	81	76	71	67
9	15	84	79	73	69
9	16	87	82	76	72

Table 12. contd.

1-tail		5%	2½%	1%	½%
2-tail		10%	5%	2%	1%
m	*n*				
9	17	90	84	78	74
9	18	93	87	80	76
9	19	96	90	83	78
9	20	99	93	85	81
9	21	102	95	88	83
9	22	105	98	90	85
9	23	108	101	93	88
9	24	111	104	95	90
9	25	114	107	98	92
10	10	82	78	74	71
10	11	86	81	77	73
10	12	89	84	79	76
10	13	92	88	82	79
10	14	96	91	85	81
10	15	99	94	88	84
10	16	103	97	91	86
10	17	106	100	93	89
10	18	110	103	96	92
10	19	113	107	99	94
10	20	117	110	102	97
10	21	120	113	105	99
10	22	123	116	108	102
10	23	127	119	110	105
10	24	130	122	113	107
10	25	134	126	116	110
11	11	100	96	91	87
11	12	104	99	94	90
11	13	108	103	97	93
11	14	112	106	100	96
11	15	116	110	103	99
11	16	120	113	107	102
11	17	123	117	110	105
11	18	127	121	113	108
11	19	131	124	116	111
11	20	135	128	119	114
11	21	139	131	123	117
11	22	143	135	126	120
11	23	147	139	129	123
11	24	151	142	132	126
11	25	155	146	136	129

1-tail		5%	2½%	1%	½%
2-tail		10%	5%	2%	1%
m	*n*				
12	12	120	115	109	105
12	13	125	119	113	109
12	14	129	123	116	112
12	15	133	127	120	115
12	16	138	131	124	119
12	17	142	135	127	122
12	18	146	139	131	125
12	19	150	143	134	129
12	20	155	147	138	132
12	21	159	151	142	136
12	22	163	155	145	139
12	23	168	159	149	142
12	24	172	163	153	146
12	25	176	167	156	149
13	13	142	136	130	125
13	14	147	141	134	129
13	15	152	145	138	133
13	16	156	150	142	136
13	17	161	154	146	140
13	18	166	158	150	144
13	19	171	163	154	148
13	20	175	167	158	151
13	21	180	171	162	155
13	22	185	176	166	159
13	23	189	180	170	163
13	24	194	185	174	166
13	25	199	189	178	170
14	14	166	160	152	147
14	15	171	164	156	151
14	16	176	169	161	155
14	17	182	174	165	159
14	18	187	179	170	163
14	19	192	183	174	168
14	20	197	188	178	172
14	21	202	193	183	176
14	22	207	198	187	180
14	23	212	203	192	184
14	24	218	207	196	188
14	25	223	212	200	192

Table 12. contd.

1-tail		5%	2½%	1%	½%
2-tail		10%	5%	2%	1%
m	n				
15	15	192	184	176	171
15	16	197	190	181	175
15	17	203	195	186	180
15	18	208	200	190	184
15	19	214	205	195	189
15	20	220	210	200	193
15	21	225	216	205	198
15	22	231	221	210	202
15	23	236	226	214	207
15	24	242	231	219	211
15	25	248	237	224	216
16	16	219	211	202	196
16	17	225	217	207	201
16	18	231	222	212	206
16	19	237	228	218	210
16	20	243	234	223	215
16	21	249	239	228	220
16	22	255	245	233	225
16	23	261	251	238	230
16	24	267	256	244	235
16	25	273	262	249	240
17	17	249	240	230	223
17	18	255	246	235	228
17	19	262	252	241	234
17	20	268	258	246	239
17	21	274	264	252	244
17	22	281	270	258	249
17	23	287	276	263	255
17	24	294	282	269	260
17	25	300	288	275	265
18	18	280	270	259	252
18	19	287	277	265	258
18	20	294	283	271	263
18	21	301	290	277	269

1-tail		5%	2½%	1%	½%
2-tail		10%	5%	2%	1%
m	n				
18	22	307	296	283	275
18	23	314	303	289	280
18	24	321	309	295	286
18	25	328	316	301	292
19	19	313	303	291	283
19	20	320	309	297	289
19	21	328	316	303	295
19	22	335	323	310	301
19	23	342	330	316	307
19	24	350	337	323	313
19	25	357	344	329	319
20	20	348	337	324	315
20	21	356	344	331	322
20	22	364	351	337	328
20	23	371	359	344	335
20	24	379	366	351	341
20	25	387	373	358	348
21	21	385	373	359	349
21	22	393	381	366	356
21	23	401	388	373	363
21	24	410	396	381	370
21	25	418	404	388	377
22	22	424	411	396	386
22	23	432	419	403	393
22	24	441	427	411	400
22	25	450	435	419	408
23	23	465	451	434	424
23	24	474	459	443	431
23	25	483	468	451	439
24	24	507	492	475	464
24	25	517	501	484	472
25	25	552	536	517	505

For larger values of m, n it is usually adequate to use a Normal approximation with continuity correction, with mean $\frac{1}{2}mn + \frac{1}{2}mn(m+1)$ and variance $\frac{1}{12}mn(m+n+1)$.

Table 13. Critical values for the Wilcoxon Single Sample and Paired Sample tests

1-tail	5%	2½%	1%	½%	1-tail	5%	2½%	1%	½%
2-tail	10%	5%	2%	1%	2-tail	10%	5%	2%	1%
n					n				
					26	110	98	84	75
2	–	–	–	–	27	119	107	92	83
3	–	–	–	–	28	130	116	101	91
4	–	–	–	–	29	140	126	110	100
5	0	–	–	–	30	151	137	120	109
6	2	0	–	–	31	163	147	130	118
7	3	2	0	–	32	175	159	140	128
8	5	3	1	0	33	187	170	151	138
9	8	5	3	1	34	200	182	162	148
10	10	8	5	3	35	213	195	173	159
11	13	10	7	5	36	227	208	185	171
12	17	13	9	7	37	241	221	198	182
13	21	17	12	9	38	256	235	211	194
14	25	21	15	12	39	271	249	224	207
15	30	25	19	15	40	286	264	238	220
16	35	29	23	19	41	302	279	252	233
17	41	34	27	23	42	319	294	266	247
18	47	40	32	27	43	336	310	281	261
19	53	46	37	32	44	353	327	296	276
20	60	52	43	37	45	371	343	312	291
21	67	58	49	42	46	389	361	328	307
22	75	65	55	48	47	407	378	345	322
23	83	73	62	54	48	426	396	362	339
24	91	81	69	61	49	446	415	379	355
25	100	89	76	68	50	466	434	397	373

For larger values of n, the Normal approximation with mean $\frac{n(n+1)}{4}$, variance $\frac{n(n+1)(2n+1)}{24}$ should be used for $T=\min[P,Q]$.

Table 14. Critical values for the Kolmogorov-Smirnov goodness-of-fit test (for completely specified distributions)

1-tail	5%	2½%	1%	½%
2-tail	10%	5%	2%	1%
n				
1	0.9500	0.9750	0.9900	0.9950
2	0.7764	0.8419	0.9000	0.9293
3	0.6360	0.7076	0.7846	0.8290
4	0.5652	0.6239	0.6889	0.7342
5	0.5094	0.5633	0.6272	0.6685
6	0.4680	0.5193	0.5774	0.6166
7	0.4361	0.4834	0.5384	0.5758
8	0.4096	0.4543	0.5065	0.5418
9	0.3875	0.4300	0.4796	0.5133
10	0.3687	0.4092	0.4566	0.4889
11	0.3524	0.3912	0.4367	0.4677
12	0.3382	0.3754	0.4162	0.4490
13	0.3255	0.3614	0.4036	0.4325
14	0.3142	0.3489	0.3897	0.4176
15	0.3040	0.3376	0.3771	0.4042
16	0.2947	0.3273	0.3657	0.3920
17	0.2863	0.3180	0.3553	0.3809
18	0.2785	0.3094	0.3457	0.3706
19	0.2714	0.3014	0.3369	0.3612
20	0.2647	0.2941	0.3287	0.3524
21	0.2586	0.2872	0.3210	0.3443
22	0.2528	0.2809	0.3139	0.3367
23	0.2475	0.2749	0.3073	0.3295
24	0.2424	0.2693	0.3010	0.3229
25	0.2377	0.2640	0.2952	0.3166
26	0.2332	0.2591	0.2896	0.3106
27	0.2290	0.2544	0.2844	0.3050
28	0.2250	0.2499	0.2794	0.2997
29	0.2212	0.2457	0.2747	0.2947
30	0.2176	0.2417	0.2702	0.2899

1-tail	5%	2½%	1%	½%
2-tail	10%	5%	2%	1%
n				
31	0.2141	0.2379	0.2660	0.2853
32	0.2108	0.2342	0.2619	0.2809
33	0.2077	0.2308	0.2580	0.2768
34	0.2047	0.2274	0.2543	0.2728
35	0.2018	0.2242	0.2507	0.2690
36	0.1991	0.2212	0.2473	0.2653
37	0.1965	0.2183	0.2440	0.2618
38	0.1939	0.2154	0.2409	0.2584
39	0.1915	0.2127	0.2379	0.2552
40	0.1891	0.2101	0.2349	0.2521
41	0.1869	0.2076	0.2321	0.2490
42	0.1847	0.2052	0.2294	0.2461
43	0.1826	0.2028	0.2268	0.2433
44	0.1805	0.2006	0.2243	0.2406
45	0.1786	0.1984	0.2218	0.2380
46	0.1767	0.1963	0.2194	0.2354
47	0.1748	0.1942	0.2171	0.2330
48	0.1730	0.1922	0.2149	0.2306
49	0.1713	0.1903	0.2128	0.2283
50	0.1696	0.1884	0.2107	0.2260
55	0.1619	0.1798	0.2011	0.2157
60	0.1551	0.1723	0.1927	0.2067
65	0.1491	0.1657	0.1853	0.1988
70	0.1438	0.1597	0.1786	0.1917
75	0.1390	0.1544	0.1727	0.1853
80	0.1347	0.1496	0.1673	0.1795
85	0.1307	0.1452	0.1624	0.1742
90	0.1271	0.1412	0.1579	0.1694
95	0.1238	0.1375	0.1537	0.1649
100	0.1207	0.1340	0.1499	0.1608

Table 15. Critical values for the Kolmogorov-Smirnov test for Normality

1-tail	5%	2½%	1%	½%
2-tail	10%	5%	2%	1%
n				
1	–	–	–	–
2	–	–	–	–
3	0.3666	0.3758	0.3812	0.3830
4	0.3453	0.3753	0.4007	0.4131
5	0.3189	0.3431	0.3755	0.3970
6	0.2972	0.3234	0.3523	0.3708
7	0.2802	0.3043	0.3321	0.3509
8	0.2652	0.2880	0.3150	0.3332
9	0.2523	0.2741	0.2999	0.3174
10	0.2411	0.2619	0.2869	0.3037
11	0.2312	0.2514	0.2754	0.2916
12	0.2225	0.2420	0.2651	0.2810
13	0.2148	0.2336	0.2559	0.2714
14	0.2077	0.2261	0.2476	0.2627
15	0.2013	0.2192	0.2401	0.2549
16	0.1954	0.2129	0.2332	0.2476
17	0.1901	0.2071	0.2270	0.2410
18	0.1852	0.2017	0.2212	0.2349
19	0.1807	0.1968	0.2158	0.2292
20	0.1765	0.1921	0.2107	0.2238
21	0.1725	0.1878	0.2060	0.2188
22	0.1688	0.1838	0.2015	0.2141
23	0.1653	0.1800	0.1974	0.2097
24	0.1620	0.1764	0.1936	0.2056
25	0.1589	0.1730	0.1899	0.2018
26	0.1560	0.1699	0.1865	0.1981
27	0.1533	0.1670	0.1833	0.1947
28	0.1507	0.1642	0.1802	0.1915
29	0.1483	0.1615	0.1773	0.1884
30	0.1460	0.1589	0.1746	0.1855

1-tail	5%	2½%	1%	½%
2-tail	10%	5%	2%	1%
n				
31	0.1437	0.1565	0.1719	0.1827
32	0.1416	0.1542	0.1693	0.1800
33	0.1395	0.1519	0.1669	0.1774
34	0.1375	0.1498	0.1645	0.1749
35	0.1356	0.1478	0.1622	0.1725
36	0.1338	0.1458	0.1601	0.1702
37	0.1321	0.1439	0.1580	0.1680
38	0.1304	0.1421	0.1560	0.1659
39	0.1288	0.1403	0.1540	0.1638
40	0.1272	0.1386	0.1522	0.1618
41	0.1257	0.1370	0.1504	0.1599
42	0.1243	0.1354	0.1487	0.1581
43	0.1229	0.1339	0.1470	0.1563
44	0.1216	0.1325	0.1452	0.1546
45	0.1203	0.1311	0.1438	0.1530
46	0.1190	0.1297	0.1423	0.1514
47	0.1178	0.1284	0.1409	0.1498
48	0.1166	0.1271	0.1394	0.1483
49	0.1155	0.1258	0.1380	0.1468
50	0.1144	0.1246	0.1367	0.1454
55	0.1092	0.1190	0.1306	0.1389
60	0.1048	0.1142	0.1253	0.1332
65	0.1008	0.1098	0.1205	0.1281
70	0.0972	0.1060	0.1163	0.1236
75	0.0940	0.1025	0.1125	0.1195
80	0.0911	0.0993	0.1090	0.1158
85	0.0885	0.0964	0.1059	0.1125
90	0.0861	0.0938	0.1030	0.1094
95	0.0838	0.0913	0.1003	0.1065
100	0.0817	0.0890	0.0978	0.1039

Table 16. Critical values for the Kolmogorov-Smirnov two-sample test

1-tail		5%	2½%	1%	½%
2-tail		10%	5%	2%	1%
n_1	n_2				
2	2	–	–	–	–
2	3	–	–	–	–
2	4	–	–	–	–
2	5	10	–	–	–
2	6	12	–	–	–
2	7	14	–	–	–
2	8	16	16	–	–
2	9	18	18	–	–
2	10	18	20	–	–
2	11	20	22	–	–
2	12	22	24	–	–
2	13	24	26	26	–
2	14	24	26	28	–
2	15	26	28	30	–
2	16	28	30	32	–
2	17	30	32	34	–
2	18	32	34	36	–
2	19	32	36	38	38
2	20	34	38	40	40
2	21	36	38	42	42
2	22	38	40	44	44
2	23	38	42	44	46
2	24	40	44	46	48
2	25	42	46	48	50
3	3	9	–	–	–
3	4	12	–	–	–
3	5	15	15	–	–
3	6	15	18	–	–
3	7	18	21	21	–
3	8	21	21	24	–
3	9	21	24	27	27
3	10	24	27	30	30
3	11	27	30	33	33
3	12	27	30	33	36
3	13	30	33	36	39
3	14	33	36	39	42
3	15	33	36	42	42
3	16	36	39	45	45
3	17	36	42	45	48
3	18	39	45	48	51
3	19	42	45	51	54

1-tail		5%	2½%	1%	½%
2-tail		10%	5%	2%	1%
n_1	n_2				
3	20	42	48	54	57
3	21	45	51	54	57
3	22	48	51	57	60
3	23	48	54	60	63
3	24	51	57	63	66
3	25	54	60	66	69
4	4	16	16	–	–
4	5	16	20	20	–
4	6	18	20	24	24
4	7	21	24	28	28
4	8	24	28	32	32
4	9	27	28	32	36
4	10	28	30	36	36
4	11	29	33	40	40
4	12	36	36	40	44
4	13	35	39	44	48
4	14	38	42	48	48
4	15	40	44	48	52
4	16	44	48	52	56
4	17	44	48	56	60
4	18	46	50	56	60
4	19	49	53	57	64
4	20	52	60	64	68
4	21	52	59	64	72
4	22	56	62	66	72
4	23	57	64	69	76
4	24	60	68	76	80
4	25	63	68	75	84
5	5	20	25	25	25
5	6	24	24	30	30
5	7	25	28	30	35
5	8	27	30	35	35
5	9	30	35	36	40
5	10	35	40	40	45
5	11	35	39	44	45
5	12	36	43	48	50
5	13	40	45	50	52
5	14	42	46	51	56
5	15	50	55	60	60
5	16	48	54	59	64
5	17	50	55	63	68

Table 16. contd.

1-tail		5%	2½%	1%	½%
2-tail		10%	5%	2%	1%
n_1	n_2				
5	18	52	60	65	70
5	19	56	61	70	71
5	20	60	65	75	80
5	21	60	69	75	80
5	22	63	70	78	83
5	23	65	72	82	87
5	24	67	76	85	90
5	25	75	80	90	95
6	6	30	30	36	36
6	7	28	30	35	36
6	8	30	34	40	40
6	9	33	39	42	45
6	10	36	40	44	48
6	11	38	43	49	54
6	12	48	48	54	60
6	13	46	52	54	60
6	14	48	54	60	64
6	15	51	57	63	69
6	16	54	60	66	72
6	17	56	62	68	73
6	18	66	72	78	84
6	19	64	70	77	83
6	20	66	72	80	88
6	21	69	75	84	90
6	22	70	78	88	92
6	23	73	80	91	97
6	24	78	90	96	102
6	25	78	88	96	107
7	7	35	42	42	42
7	8	34	40	42	48
7	9	36	42	47	49
7	10	40	46	50	53
7	11	44	48	55	59
7	12	46	53	58	60
7	13	50	56	63	65
7	14	56	63	70	77
7	15	56	62	70	75
7	16	59	64	73	77
7	17	61	68	77	84
7	18	65	72	83	87
7	19	69	76	86	91

1-tail		5%	2½%	1%	½%
2-tail		10%	5%	2%	1%
n_1	n_2				
7	20	72	79	91	93
7	21	77	91	98	105
7	22	77	84	97	103
7	23	80	89	101	108
7	24	84	92	105	112
7	25	86	97	108	115
8	8	40	48	48	56
8	9	40	46	54	55
8	10	44	48	56	60
8	11	48	53	61	64
8	12	52	60	64	68
8	13	54	62	67	72
8	14	58	64	72	76
8	15	60	67	75	81
8	16	72	80	88	88
8	17	68	77	85	88
8	18	72	80	88	94
8	19	74	82	93	98
8	20	80	88	100	104
8	21	81	89	102	107
8	22	84	94	106	112
8	23	89	98	107	115
8	24	96	104	120	128
8	25	95	104	118	125
9	9	54	54	63	63
9	10	50	53	61	63
9	11	52	59	63	70
9	12	57	63	69	75
9	13	59	65	73	78
9	14	63	70	80	84
9	15	69	75	84	90
9	16	69	78	87	94
9	17	74	82	92	99
9	18	81	90	99	108
9	19	80	89	99	107
9	20	84	93	104	111
9	21	90	99	111	117
9	22	91	101	113	122
9	23	94	106	117	126
9	24	99	111	123	132
9	25	101	114	125	135

Table 16. contd.

1-tail		5%	2½%	1%	½%
2-tail		10%	5%	2%	1%
n_1	n_2				
10	10	60	70	70	80
10	11	57	60	69	77
10	12	60	66	74	80
10	13	64	70	78	84
10	14	68	74	84	90
10	15	75	80	90	100
10	16	76	84	94	100
10	17	79	89	99	106
10	18	82	92	104	108
10	19	85	94	104	113
10	20	100	110	120	130
10	21	95	105	118	126
10	22	98	108	120	130
10	23	101	114	127	137
10	24	106	118	130	140
10	25	110	125	140	150
11	11	66	77	88	88
11	12	64	72	77	86
11	13	67	75	86	91
11	14	73	82	90	96
11	15	76	84	95	102
11	16	80	89	100	106
11	17	85	93	104	110
11	18	88	97	108	118
11	19	92	102	114	122
11	20	96	107	118	127
11	21	101	112	124	134
11	22	110	121	143	143
11	23	108	119	132	142
11	24	111	124	139	150
11	25	117	129	143	154
12	12	72	84	96	96
12	13	71	81	92	95
12	14	78	86	94	104
12	15	84	93	102	108
12	16	88	96	108	116
12	17	90	100	112	119
12	18	96	108	120	126
12	19	99	108	121	130
12	20	104	116	128	140
12	21	108	120	132	141

1-tail		5%	2½%	1%	½%
2-tail		10%	5%	2%	1%
n_1	n_2				
12	22	110	124	138	148
12	23	113	125	138	149
12	24	132	144	156	168
12	25	120	138	153	165
13	13	91	91	104	117
13	14	78	89	102	104
13	15	87	96	107	115
13	16	91	101	112	121
13	17	96	105	118	127
13	18	99	110	123	131
13	19	104	114	130	138
13	20	108	120	135	143
13	21	113	126	140	150
13	22	117	130	143	156
13	23	120	135	152	161
13	24	125	140	155	166
13	25	131	145	160	172
14	14	98	112	112	126
14	15	92	98	111	123
14	16	96	106	120	126
14	17	100	111	125	134
14	18	104	116	130	140
14	19	110	121	135	148
14	20	114	126	142	152
14	21	126	140	154	161
14	22	124	138	152	164
14	23	127	142	159	170
14	24	132	146	164	176
14	25	136	150	169	182
15	15	105	120	135	135
15	16	101	114	120	133
15	17	105	116	131	142
15	18	111	123	138	147
15	19	114	127	142	152
15	20	125	135	150	160
15	21	126	138	156	168
15	22	130	144	160	173
15	23	134	149	165	179
15	24	141	156	174	186
15	25	145	160	180	195

Table 16. contd.

1-tail		5%	2½%	1%	½%
2-tail		10%	5%	2%	1%
n_1	n_2				
16	16	112	128	144	160
16	17	109	124	139	143
16	18	116	128	142	154
16	19	120	133	151	160
16	20	128	140	156	168
16	21	130	145	162	173
16	22	136	150	168	180
16	23	141	157	175	187
16	24	152	168	184	200
16	25	149	167	186	199
17	17	136	136	153	170
17	18	118	133	150	164
17	19	126	141	158	166
17	20	130	146	163	175
17	21	136	151	168	180
17	22	142	157	176	187
17	23	146	163	181	196
17	24	151	168	187	203
17	25	156	173	196	207
18	18	144	162	180	180
18	19	133	142	160	176
18	20	136	152	170	182
18	21	144	159	177	189
18	22	148	164	184	196
18	23	152	170	189	204
18	24	162	180	198	216
18	25	162	180	202	216
19	19	152	171	190	190
19	20	144	160	171	187
19	21	147	163	184	199
19	22	152	169	190	204
19	23	159	177	197	209
19	24	164	183	204	218
19	25	168	187	211	224
20	20	160	180	200	220
20	21	154	173	193	199
20	22	160	176	196	212
20	23	164	184	205	219
20	24	172	192	212	228
20	25	180	200	220	235

1-tail		5%	2½%	1%	½%
2-tail		10%	5%	2%	1%
n_1	n_2				
21	21	168	189	210	231
21	22	163	183	205	223
21	23	171	189	213	227
21	24	177	198	222	237
21	25	182	202	225	244
22	22	198	198	242	242
22	23	173	194	217	237
22	24	182	204	228	242
22	25	189	209	234	250
23	23	207	230	253	253
23	24	183	205	228	249
23	25	195	216	243	262
24	24	216	240	264	288
24	25	204	225	254	262
25	25	225	250	275	300
26	26	234	260	286	312
27	27	243	270	324	324
28	28	280	308	336	364
29	29	290	319	348	377
30	30	300	330	360	390
31	31	310	341	372	403
32	32	320	352	416	416
33	33	330	396	429	462
34	34	374	408	442	476
35	35	385	420	455	490
36	36	396	432	468	504
37	37	407	444	518	518
38	38	418	456	532	570
39	39	429	468	546	585
40	40	440	520	560	600
41	41	492	533	574	615
42	42	504	546	588	630
43	43	516	559	645	688
44	44	528	572	660	704
45	45	540	585	675	720
46	46	552	644	690	736
47	47	564	658	705	752
48	48	576	672	720	768
49	49	637	686	735	833
50	50	650	700	800	850

Table 17. Random Numbers

68236	35335	71329	96803	24413
62385	36545	59305	59948	17232
64058	80195	30914	16664	50818
64822	68554	90952	64984	92295
17716	22164	05161	04412	59002
03928	22379	92325	79920	99070
11021	08533	83855	37723	77339
01830	68554	86787	90447	54796
36782	73208	93548	77405	58355
58158	45059	83980	40176	40737
91239	10532	27993	11516	61327
27073	98804	60544	12133	01422
81501	00633	62681	84319	03374
64374	26598	54466	94768	19144
29896	26739	30871	29795	13472
38996	72151	65746	16513	62796
73936	81751	00149	99126	23117
18795	93118	81105	10007	49807
76816	99822	92314	45035	43490
12091	60413	90467	42457	50490
41538	19059	69055	94355	84262
12909	04950	14986	08205	53582
49185	94608	87317	37725	66450
37771	48526	14939	32848	77677
22532	13814	69092	78342	37774
60132	24386	10989	54346	41531
23784	56693	45902	33406	53867
03081	20189	77226	89923	67301
51273	64049	19919	45518	43243
03281	40214	60679	68712	71636

Internet links

Finding statistics resources on the internet

The internet has a vast amount of information on statistics available to academics, professional researchers, and laypeople alike. Finding reliable and free data should not be difficult although a few points need to be borne in mind. The material should be up to date. Small organizations and departments within academic institutions sometimes encounter funding difficulties and are unable to continue with their researches. Make sure to look at the 'Last updated' section of the main website page before using any data. Try clicking on the links to make sure that they have been maintained properly and do not result in error messages. Ideally information should be obtained from websites run by universities, research institutes, and other reputable organizations. Websites maintained by individuals may not be up to date and comprehensive. It is also possible that the prejudices of those maintaining the websites will be reflected in the content and list of links.

General Resources

Current Index to Statistics

www.statindex.org/CIS/query

A bibliographic index to publications in statistics and related fields. References are drawn from over 160 core journals that are fully indexed, non-core journals from which articles are selected that have statistical content, proceedings and edited books, and other sources. The Current Index to Statistics is a joint venture of the American Statistical Association and the Institute of Mathematical Statistics.

Probability Abstract Service

www.economia.unimi.it/PAS/

An archive of research article abstracts which publishes a bi-monthly newsletter.

The WWW Virtual Library: Statistics

www.stat.ufl.edu/vlib/statistics.html

Catalog of statistics resources includes university departments, government departments, research groups, journals, software, and newsgroups.

MathWorld

http://mathworld.wolfram.com

Comprehensive and interactive mathematics encyclopedia intended for students, educators, math enthusiasts, and researchers assembled over more than a decade by internet encyclopedist Eric W. Weisstein with assistance from the mathematics and internet communities. MathWorld is hosted and sponsored by Wolfram Research, Inc., makers of Mathematica. Links to mathematics and science sites owned by Wolfram Resources.

Probability Web

www.mathcs.carleton.edu/probweb/

A collection of probability resources on the Web designed to be especially helpful to researchers, teachers, and people in the probability community.

Statistical Science Web

www.statsci.org/

A window to statistical science and bioinformatics on the web, with special attention to Australia. StatSci.org consists of two parts: resources and directory.

Online Calculators and Software:

Statistical Java

www.stat.vt.edu/%7Esundar/java/applets/

An interactive environment for teaching statistics from Virginia Polytechnic Institute's Department of Statistics.

Statistics Calculator
www.cebm.utoronto.ca/practise/ca/statscal/
Calculator offered by the Centre for Evidence-Based Medicine at the University of Toronto.

Online Tutorials and Textbooks

The Statistics Homepage
www.statsoft.com/textbook/stathome.html
Award-winning online statistics text containing techniques often required by science, engineering, medical, finance, and business students.

Basic Principles of Statistical Analysis
http://duke.usask.ca/~rbaker/stats.html
Online textbook maintained by Professor Bob Baker, University of Saskatchewan.

A Compendium of Common Probability Distributions
www.causascientia.org/math_stat/Dists/Compendium.pdf
Fifty-six probability distributions and their properties, including PDF, CDF, Moments, Random-Variate Generation, and so on.

Introductory Statistics: Concepts, Models, and Applications
www.psychstat.smsu.edu/introbook/sbk00.htm
Online course textbook authored by David W. Stockburger at Southwest Missouri State University.

Statistics.com
www.statistics.com/
A site primarily offering online courses but also containing links to online textbooks, glossaries and other reference material, commercial and free statistical software, and sites of statistical data.

History

History of Statistics
www.york.ac.uk/depts/maths/histstat/welcome.htm
Compilation of online materials accessed for a history of statistics course offered by the University of York's Department of Mathematics.

Historia Matematica
www.chasque.net/jgc/history/MH4.htm
The purpose of this forum is to provide a virtual environment for scholarly discussion of the history of mathematics (in a broad sense), amongst professionals, and nonprofessionals with a serious interest in the field.

MacTutor History of Mathematics
www-groups.dcs.st-and.ac.uk/~history
Offers mathematics history, chronologies, and more than 1600 biographies of mathematicians. The MacTutor History of Mathematics archive was created and is maintained by John O'Connor and Edmund F. Robertson at the University of St. Andrews, Scotland.

The Cornell University Library: Historical Mathematics Monographs
http://historical.library.cornell.edu/math/
A collection of selected monographs with expired copyrights chosen from the mathematics field.

British Society for the History of Mathematics
www.dcs.warwick.ac.uk/bshm/index.html
The aims of the British Society for the History of Mathematics are to promote research into the history of mathematics and its use at all levels of mathematics education.

The Mathematical Museum: The History Wing
www.math-net.org/links/show?collection=math.museum.hist
The 'History' wing of The Mathematical Museum contains pointers to selected exhibitions,

hyperbooks, information systems, museums, and pages related to the history of mathematics and adjacent fields.

Institute for History and Foundations of Science
www.phys.uu.nl/~wwwgrnsl/indexi.html
The Institute for History and Foundations of Science (Utrecht University) is part of the Faculty of Physics and Astronomy. The Institute consists of two distinct sections: the History of Science Section and the Foundations of Science Section.

Careers

American Statistical Association – Careers
www.amstat.org/careers
Information regarding careers in statistics.

The Mathematical Association of America: Careers in Mathematics
www.maa.org/students/career.html
FAQs regarding careers in mathematics.

Mathematical Sciences Career Information
www.ams.org/careers/
The American Mathematical Society, the Mathematical Association of America, and the Society for Industrial and Applied Mathematics are dedicated to providing career information and services to the mathematics community.

Organizations

International Statistical Institute
http://isi.cbs.nl/
The International Statistical Institute (ISI) is one of the oldest international scientific associations functioning in the modern world. It is an autonomous society, which seeks to develop and improve statistical methods and their application through the promotion of international activity and co-operation.

UNESCO Institute for Statistics
www.uis.unesco.org
The UNESCO Institute for Statistics is the statistical branch of the United Nations Organization for Education, Science and Culture (UNESCO).

American Statistical Association
www.amstat.org
The American Statistical Association is a scientific and educational society founded in 1839 with the following mission: to promote excellence in the application of statistical science across the wealth of human endeavor.

Institute of Mathematical Statistics
www.imstat.org/
The IMS is an international professional and scholarly society devoted to the development, dissemination, and application of statistics and probability.

International Association for Statistics Education (IASE)
www.stat.auckland.ac.nz/~iase/
IASE seeks to improve statistics education at all levels from elementary (primary) school through to the training of professionals, and to increase the uptake of statistics education worldwide.

International Biometric Society
www.tibs.org/
The International Biometric Society is an international society promoting the development and application of statistical and mathematical theory and methods in the biosciences, including

agriculture, biomedical science and public health, ecology, environmental sciences, forestry, and allied disciplines.

International Environmetrics Society (TIES)
www.nrcse.washington.edu/ties/
TIES seeks to foster the development and use of statistical and other quantitative methods in the environmental sciences, environmental engineering, and environmental monitoring and protection.

International Society for Clinical Biostatistics
www.iscb.info/
ISCB was founded in 1978 to stimulate research into the principles and methodology used in the design and analysis of clinical research and to increase the relevance of statistical theory to the real world of clinical medicine.

Journals

American Statistician
www.amstat.org/publications/tas
Quarterly publication by the American Statistical Association contains articles organized into the following sections: Statistical Practice, General, Teacher's Corner, Statistical Computing and Graphics, Reviews of Books and Teaching Materials, and Letters.

Biometrika
http://biomet.oxfordjournals.org/
Biometrika is primarily a journal of statistics in which emphasis is placed on papers containing original theoretical contributions of direct or potential value in applications.

Canadian Journal of Statistics
http://archimede.mat.ulaval.ca/cjs/
The Canadian Journal of Statistics is an official publication of the Statistical Society of Canada. It publishes research articles of theoretical, applied, or pedagogical interest to the statistical community. In French and in English.

Journal of Agricultural, Biological and Environmental Statistics
www.amstat.org/publications/jabes/
The purpose of the *Journal of Agricultural, Biological, and Environmental Statistics (JABES)* is to contribute to the development and use of statistical methods in the agricultural sciences, the biological sciences (including biotechnology), and the environmental sciences (including those dealing with natural resources).

Data/Sources

Statistical Resources
www.lib.umich.edu/govdocs/stats.html
The Documents Center is a central reference and referral point for government information, whether local, state, federal, foreign, or international. From the University of Michigan Documents Center.

MathDL: The MAA Mathematical Sciences Digital Library
www.mathdl.org/
An online resource published by the Mathematical Association of America. The site provides resources for both teachers and students of mathematics.

arXiv.org e-Print archive
http://arxiv.org/
An e-print service in the fields of physics, mathematics, nonlinear science, computer science, and quantitative biology. The contents of arXiv conform to Cornell University academic standards. arXiv is owned, operated, and funded by Cornell University.

EULER – Your Portal to Mathematics Publications
www.emis.de/projects/EULER/
European-based virtual library for mathematics. In particular, EULER provides a world reference and delivery service, transparent to the end user and offering full coverage of the mathematics literature worldwide, including bibliographic data, peer reviews, and abstracts.

Zentralblatt MATH
www.emis.de/ZMATH/
The world's most complete and longest running abstracting and reviewing service in pure and applied mathematics. The Zentralblatt MATH Database contains more than two million entries drawn from more than 2300 serials and journals and covers the period from 1868 to the present by the recent integration of the Jahrbuch database (JFM).

Population Reference Bureau
www.prb.org/
The Population Reference Bureau's mission is to be the leader in providing timely and objective information on US and international population trends and their implications.

US Census Bureau International Data Base
www.census.gov/ipc/www/idbnew.html
The International Data Base (IDB) is a data bank containing statistical tables of demographic and socioeconomic data for 227 countries and areas of the world.

Glossaries

VIROS - Virtual Institute for Research in Official Statistics (Eurostat)
http://europa.eu.int/comm/eurostat/research/index.htm?http://europa.eu.int/en/comm/eurostat/research/isi/&1
The International Statistical Institute (ISI) has compiled a glossary of statistical terms, in a number of languages, some of which use special characters.

Statistics Homepage Glossary
www.statsoft.com/textbook/glosfra.html
Entries in the Statistical Glossary are taken from the Electronic Manual of STATISTICA and may contain elements that refer to specific features of the STATISTICA system.